Routledge Revivals

North Pacific Fisheries Management

In anticipation of the UN Conference of the Law of the Sea taking place in 1973, Dr Kasahara and Dr Burke of the University of Washington first published *North Pacific Fisheries Management* earlier that year. The conference brought fishery territories to a global stage with delegates that may not be as informed about ocean issues as those previously making decisions. Therefore the Program of International Studies of Fishery Arrangements was created to explore the management of fisheries in specific regions. This study focusses on the North Pacific region and delves into the implications of a global regime, generic problems concerning fishery management, distribution and institutions as well as alternative arrangements that can be made to make the management of fisheries smoother. This title will be of interest to students of environmental studies and policy makers.

North Pacific Fisheries Management

Hiroshi Kasahara and William Burke

First published in 1973
by Resources for the Future, Inc.

This edition first published in 2016 by Routledge
2 Park Square, Milton Park, Abingdon, Oxon, OX14 4RN
and by Routledge
711 Third Avenue, New York, NY 10017

Routledge is an imprint of the Taylor & Francis Group, an informa business

Publisher's Note
The publisher has gone to great lengths to ensure the quality of this reprint but points out that some imperfections in the original copies may be apparent.

Disclaimer
The publisher has made every effort to trace copyright holders and welcomes correspondence from those they have been unable to contact.

A Library of Congress record exists under LC control number: 74159058

ISBN 13: 978-1-138-94626-2 (hbk)
ISBN 13: 978-1-315-67088-1 (ebk)

NORTH PACIFIC FISHERIES MANAGEMENT

NORTH PACIFIC FISHERIES MANAGEMENT

Hiroshi Kasahara
and
William Burke

Paper no. 2 in a series prepared for
THE PROGRAM OF INTERNATIONAL STUDIES OF FISHERY ARRANGEMENTS
Francis T. Christy, Jr., Director

RESOURCES FOR THE FUTURE, INC.
Washington, D.C.

May 1973

RESOURCES FOR THE FUTURE, INC.
1755 Massachusetts Avenue, N.W., Washington, D.C. 20036

Resources for the Future is a nonprofit corporation for research and education in the development, conservation, and use of natural resources and the improvement of the quality of the environment. It was established in 1952 with the cooperation of the Ford Foundation. Part of the work of Resources for the Future is carried out by its resident staff; part is supported by grants to universities and other nonprofit organizations. Unless otherwise stated, interpretations and conclusions in RFF publications are those of the authors; the organization takes responsibility for the selection of significant subjects for study, the competence of the researchers, and their freedom of inquiry. The manuscript was edited by Sheila Barrows.

RFF editors: Mark Reinsberg, Vera W. Dodds, Ruth B. Haas

RFF—PISFA Paper 2. $3.00

Contents

Figures

Note: Figure 2 reprinted, by permission, from Koblentz-Mishke, Volkovinsky, and Kabanova (1970). Figures 6 and 7 are reprinted, with permission, from Kasahara (1961). Figures 8-17 are reprinted, with permission, from Kasahara (1972).

Preface

THE THIRD UNITED NATIONS CONFERENCE ON THE LAW OF THE SEA is now scheduled to open at the end of 1973 and begin its first substantive sessions early in 1974. Of the many problems that will face the conference delegates, those dealing with the management and distribution of marine fisheries are among the most important, most difficult, and least understood. Almost all coastal states have an interest in marine fisheries. Even though this interest may be rudimentary, there is still a political constituency that will have an influence on the decisions of the delegations. The problems of fisheries are difficult because many stocks of fish have migratory patterns that extend beyond the jurisdictions of single states, either paralleling the coast into the waters of neighboring states or outwards into the high seas. And with present trends towards the dissolution of the principle of the "freedom of the seas" for fishing, joint decisions on the distribution of the seas' wealth become necessary. In the past, fishery decisions have generally been made on specific problems by fishery experts. With the coming conference, however, the decisions will be made in a global arena, in the broad context of a multitude of ocean issues, and by delegates with only a partial knowledge of fishery matters.

For these reasons Resources for the Future, Inc. decided to concentrate its current ocean interests on fisheries and, with the help of a supplemental grant from the Ford Foundation, initiated the Program of International Studies of Fishery Arrangements. The objective of the program is to produce information that will facilitate the preparations for the UN Conference and contribute to the process of making decisions. The program approaches this objective by concentrating primarily on particular fisheries and fishery regions, attempting to elucidate the variety of alternatives that exist in different, real situations. It is hoped that this will produce a better understanding of the implications of the proposals for universal regimes and principles that are likely to emerge in the UN discussions. In addition, the program examines some of the generic problems of fisheries management, distribution, and institutions.

Each study provides background information on recent developments and trends and a discussion of alternative legal and institu-

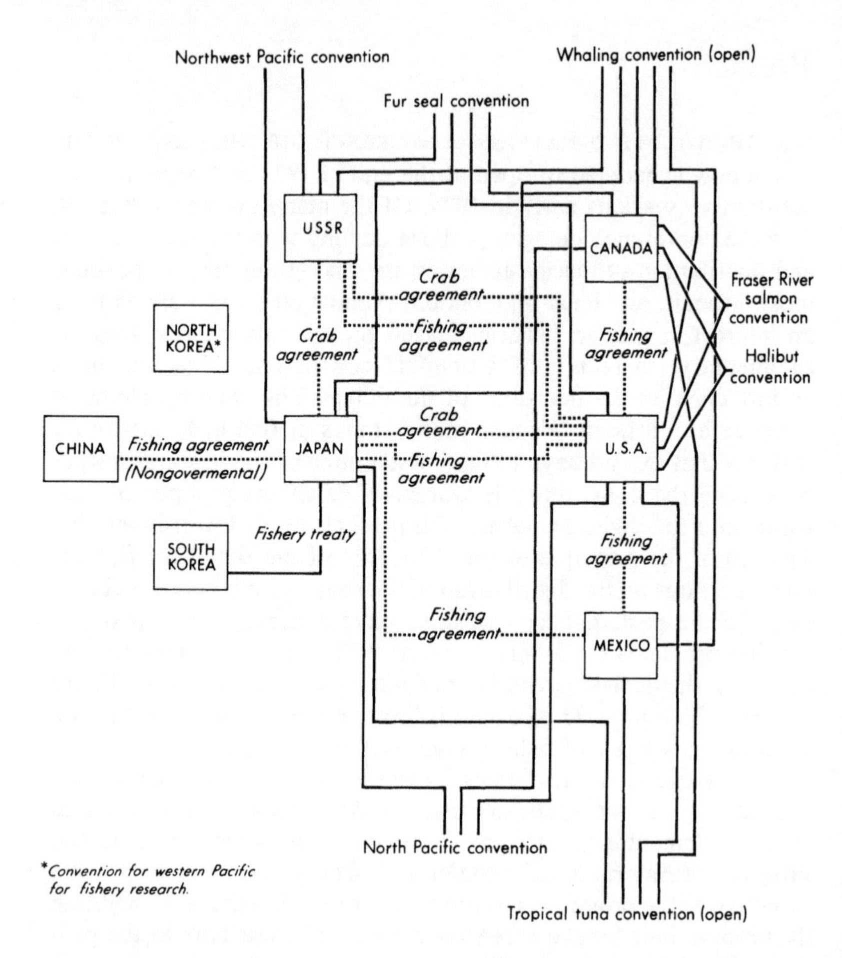

Fishery conventions and agreements in the North Pacific (as of October 1970).

tional arrangements for the resolution of the problems. The studies attempt to raise questions and suggest approaches that will be helpful to the decision makers, rather than to recommend specific courses of action. Every effort has been made to ensure that each study has been prepared from a non-national perspective and that it has taken into account all responsible points of view and interests. The studies are freely available to all delegates at the UN preparatory sessions as well as at the conference. In addition, the program seeks opportunities for full and free discussion of the studies with interested persons. Eventually, all of the separate studies may be put together in a single volume in order to meet the anticipated continuing demand for information on fishery arrangements. Comments and criticisms of the individual papers are solicited and will be considered for publication in such a volume.

Studies in the Program include:

An overview of fishery arrangements;
North Pacific fisheries management;
East Central Atlantic fisheries;
Indian Ocean fisheries;
World tuna fisheries;
Alternative international institutions.

Each of the regional studies in the program illustrates a different kind of situation. The present study deals with an area in which the relatively few participants have had a long history of fishing and considerable experience in international negotiations. As shown in the accompanying chart, fisheries in the North Pacific have been subjected to a highly complex overlay of bilateral and multilateral agreements and are operating under a wide variety of commissions. As a result, the problems for the future lie not so much in the adoption of new arrangements and agreements as in the adaptation of the old. Thus, the study concentrates primarily on institutions that can help to reduce some of the complexity and increase the flexibility, so that the changes taking place can be accommodated with greater ease.

Dr. Hiroshi Kasahara was Professor of Fisheries, College of Fisheries at the University of Washington when this study was written. Before that he was in charge of fisheries for the United Nations Development Program. Dr. William T. Burke, in his capacity as Professor of Law at the University of Washington, has undertaken many studies of international fisheries and other ocean problems. He was coauthor, with Professor Myres McDougal, of *The Public Order of the Oceans* — the most comprehensive recent treatment of the law of the sea.

FRANCIS T. CHRISTY, JR., Director
RFF Program of International
Studies of Fishery Arrangements

Introduction

THE SPECIFIC NEEDS AND PROBLEMS of international fishery management differ from region to region around the world but are becoming more urgent and complex in every geographic locale. Difficulties in managing living marine resources are not new; exploitation of these resources has intensified markedly over the past two decades and the resulting problems have been well-known to national and international officials for almost as long.

These more familiar ocean resource management problems are being joined by the many other issues of ocean development, management, and control now under discussion in the United Nations preparatory to a third Conference on the Law of the Sea. These issues will be examined at the proposed conference in order to reach agreement upon international legal prescriptions and procedures by which the ocean and its resources may be managed and benefits divided or shared. More specifically, the states of the world will apparently seek to provide a general arrangement whereby management authority over all ocean fishery resources is allocated among states and international entities. In addition, it is expected that states will attempt to provide for an allocation of fishery resources or the benefits of their exploitation.

This book attempts to place the fishery management problems of the North Pacific Ocean in the context both of their historical treatment and of the unfolding future of exploitation and management, including the impact of the United Nations Conference on the Law of the Sea.[1] This region includes some of the most highly productive fisheries, as well as the strongest fishing states, in the world. It has been noted for its rich fishery resources for many decades and controversies over them extend back into the latter part of the nineteenth century. The states of the region have had long experience in the resolution of these controversies and have employed many different institutional and substantive arrangements in their disposition. Chapters 1 and 2 examine the status of exploitation of North Pacific fishery resources and the interests of Pacific rim states in such resources, and mention briefly the long record of controversy over them, with emphasis on the specific responses by states to resolve their disputes.

Fifteen relevant international agreements are currently applicable to North Pacific fishery resources and problems. Despite, and partly because of, this surfeit of agreements, critical management problems are still outstanding and more are likely to develop in the foreseeable future. The vast bulk of North Pacific fisheries is still outside any international regulatory arrangement, despite the many current agreements, and at present no effective institutional mechanism exists even for detecting, let alone resolving, emerging problems. Chapter 3 identifies and briefly discusses the existing and expected fishery management problems of the North Pacific Ocean.

A number of possible approaches might be made to resolve these fishery problems in the North Pacific, ranging from reliance on unilateral actions, to continuation of present ad hoc methods, to arrangements reached at the third LOS Conference. A variety of substantive alternatives are now being examined and debated in the preparations for this conference. Generally these various proposals are most directly concerned with such policy issues as the area of ocean fishing to be subjected to coastal control, the degree and kind of authority to be exercised by coastal states, and, most important, the allocation of benefits of fishing among coastal and other participants. The key question is whether and how one or another choice on these issues would resolve North Pacific problems. Chapter 4 considers various alternatives and concludes that none can hope to avoid continuing negotiations for coping with the management difficulties that can be expected to persist.

It is inevitable that the states in the North Pacific will continue to confront management problems over some years, even assuming that the third LOS Conference were to reach effective agreement(s) on fisheries. If, as is likely, the conference does not reach effective agreement, the controversies will probably be sharper and more intense. In sum, whatever the future brings, there will be a need for a better institutional structure for making decisions about fisheries in the North Pacific. Chapter 5 considers the means for transforming the multitude of existing arrangements into a new and comprehensive institution for North Pacific fisheries management and examines the composition and functioning of such an institution.

[1]Hereafter designated as the LOS Conference.

I Exploitation of Fishery Resources and State Interests in the North Pacific

OUR PRINCIPAL OBJECTIVE is to review past and future problems of international fishery management in the North Pacific Ocean and to consider alternative institutional arrangements and substantive policies that might assist in resolving such problems. First, however, some background information is necessary in order to understand the scale and complexity of fishing activities in this region.

Description of the area designated as North Pacific in this discussion presents a problem. For the present we are mainly concerned with the exploitation of living resources in waters north of the tropics, and it is difficult to draw a more clear-cut line. In practical terms, tuna fishing in tropical waters is largely excluded from present consideration. For one thing, in contrast to other stocks of interest, the distribution of tropical tuna species is continuous from the northern to the southern hemisphere. Inclusion of the tropical part of the North Pacific would also call for discussion of the complicated situation in Southeast Asia, which differs greatly from the problems elsewhere and has little bearing on those to be confronted in the North Pacific.

Description of the general oceanography of the North Pacific is unnecessary but some oceanographic features are worth mentioning since they reflect (or partially determine) the distribution of major resources, both real and potential, and are often mentioned as a possible basis for delimiting coastal authority over fisheries. The first of these factors is the distribution of the continental shelf and slope, as shown in figure 1. Practically all fishing for demersal species now occurs within the continental shelf and the upper half of the slope. Further development of fishing technology will undoubtedly make it feasible to exploit demersal species to a depth of 1,000 meters or to even greater depths, at least in some parts of the region, but this will still be generally within the outer margin

Figure 1. The northern Pacific Ocean.

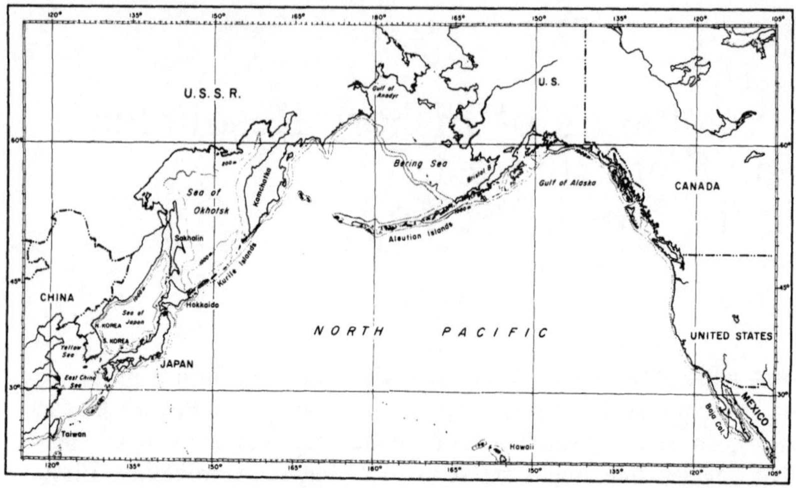

Map of the North Pacific with 200 m and 1000 m contours.

of the continental slope. Many of the important pelagic fish resources, too, occur in waters superjacent to the shelf and slope.

Shelf areas of large-to-medium geographic expanse are found from the Gulf of Alaska to the East China Sea. Particularly large areas occur in the Bering Sea, the Okhotsk Sea, and the East China Sea (including the Yellow Sea), extending in some places outward up to several hundred miles from the coast. The northern part of the Bering Sea shelf and the northern Okhotsk Sea shelf are, however, of limited importance to fisheries because of the low bottom temperatures (near or below 0° C) that prevail even in the summer. While bottom faunas in the shelf areas of the northeastern Pacific, Bering Sea, Okhotsk Sea, and Japan Sea, as well as in waters off the Kurile chain and northern Japan, are principally subarctic (or arctic in the case of the northern parts of the Bering Sea and Okhotsk Sea), the East China Sea is dominated by temperate demersal species (Kasahara 1964, p. 67).

A second oceanographic feature that is important from the point of view of fisheries is the difference in the surface temperature regime between the western and eastern North Pacific. Because of the general circulation of water in the region and the strong influence of continental climate on the Asian side, the upper zone of the western half and its adjacent seas are characterized by seasonal temperature changes that are much greater than those in the eastern half of the Pacific Ocean proper, and by sharp boundary situations resulting in marked seasonal movements and concentrations of pelagic species.

Primary production is often used as a rough indication of an abundance of animals at higher levels of the food chain. This is done for two reasons: (a) the rate at which organic carbon is produced by primary producers should reflect relative productivity of different areas, and (b) primary production can be determined fairly accurately through measurements of ^{14}C uptake. Primary production data, however, are sometimes misleading as an indication of potential yields at higher levels, particularly in areas at high latitudes where phytoplankton grow very rapidly for a short period of time. At the next higher level, the zooplankton, most data are expressed in terms of biomass rather than of production, and therefore cannot be directly related to yields at higher levels. Despite various limitations, however, the assumption that the areas of high primary production *generally* coincide with those abundant in exploitable living resources is supported by the actual distribution of fishing effort and catches.

Figure 2 shows that the areas of high primary production are along the rim of the ocean and in areas of divergence associated with the equatorial current system. Except for tunas, most of the good fishing grounds are also along the rim of the ocean. Waters of the equatorial current system are known to contain abundant

Figure 2. Distribution of primary production in the world ocean.

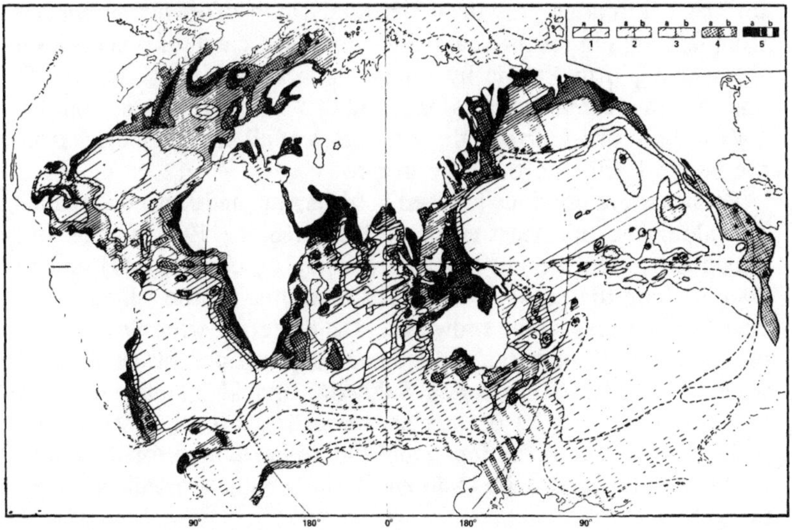

Note: **Units are in mg of C per m² per day. (1) less than 100; (2) 100-150; (3) 150-250; (4) 250-500; (5) more than 500. *a* = data from direct ^{14}C measurements; *b* = data from phytoplankton biomass, hydrogen, or oxygen saturation.**

stocks of unutilized species and to cover some of the major tuna fishing grounds. Perhaps the most important aspect of primary production data summarized in figure 2 is the existence of a vast area in the central part of the North Pacific in which abundant potential resources are not likely to be found.

EXPLOITATION OF RESOURCES

Northwestern Pacific and Bering Sea

The Bering Sea and the Northwestern Pacific—including the Okhotsk Sea, the Japan Sea, the East China and Yellow Seas, and Pacific waters off the Japanese islands—constitute one of the most intensively fished areas of the world ocean. Major fishing grounds in this general region are heavily fished by Japan, the Soviet Union, South and North Korea, the People's Republic of China, and Taiwan. In the eastern Bering Sea, the United States also exploits salmon, crabs, and halibut. Most of the stocks of commercially important demersal species, both fish and invertebrates occurring to a depth of approximately 500 meters, are either fully exploited or even overfished, so that in general their catches will not increase as fishing is further intensified. A possible exception is pollack (*Theragra chalocogramma*), which is largely responsible for the continuous growth of the total catch of demersal species from this area in recent years. The combined catch of pollack over the whole North Pacific by Japan, the Soviet Union, and North and South Korea perhaps reached 3.5 million metric tons in 1971—about one-fourth of the total catch of all marine species from the North Pacific north of the tropics.

Pollack were previously fished in waters around Japanese islands and along the northeast coast of Korea. Since 1962, fishing has expanded to the Bering Sea (particularly its eastern half) and to waters along the coast of Kamchatka and the Kurile islands. The best fishing areas so far known are: the eastern Bering Sea; waters off southwest Kamchatka and the northernmost part of the Kuriles; waters around Hokkaido and Sakhalin; and along the northeast coast of Korea, more or less in this order. Figure 3 shows the pollack catch trends for Japan and the Soviet Union for the entire North Pacific, and a breakdown of the Japanese catch is given in figure 4.

The expansion of Japanese pollack fishing in the Bering Sea came about when the yellowfin sole (*Limanda aspera*) stock fished by the Japanese fish meal factory-ship fleet declined sharply in 1964

Figure 3. Catches of pollack by Japan, the USSR, and South Korea, 1958-70.

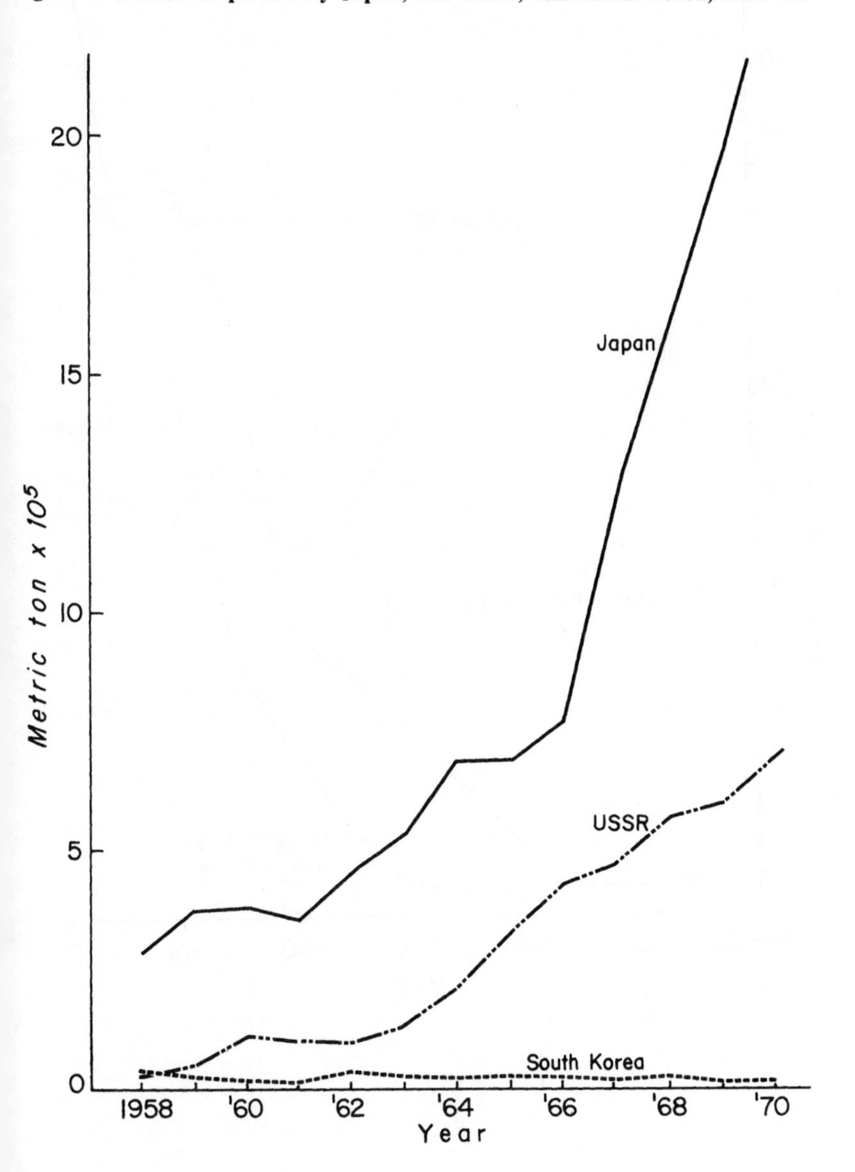

Figure 4. Catches of pollack by different Japanese fisheries, 1959-70.

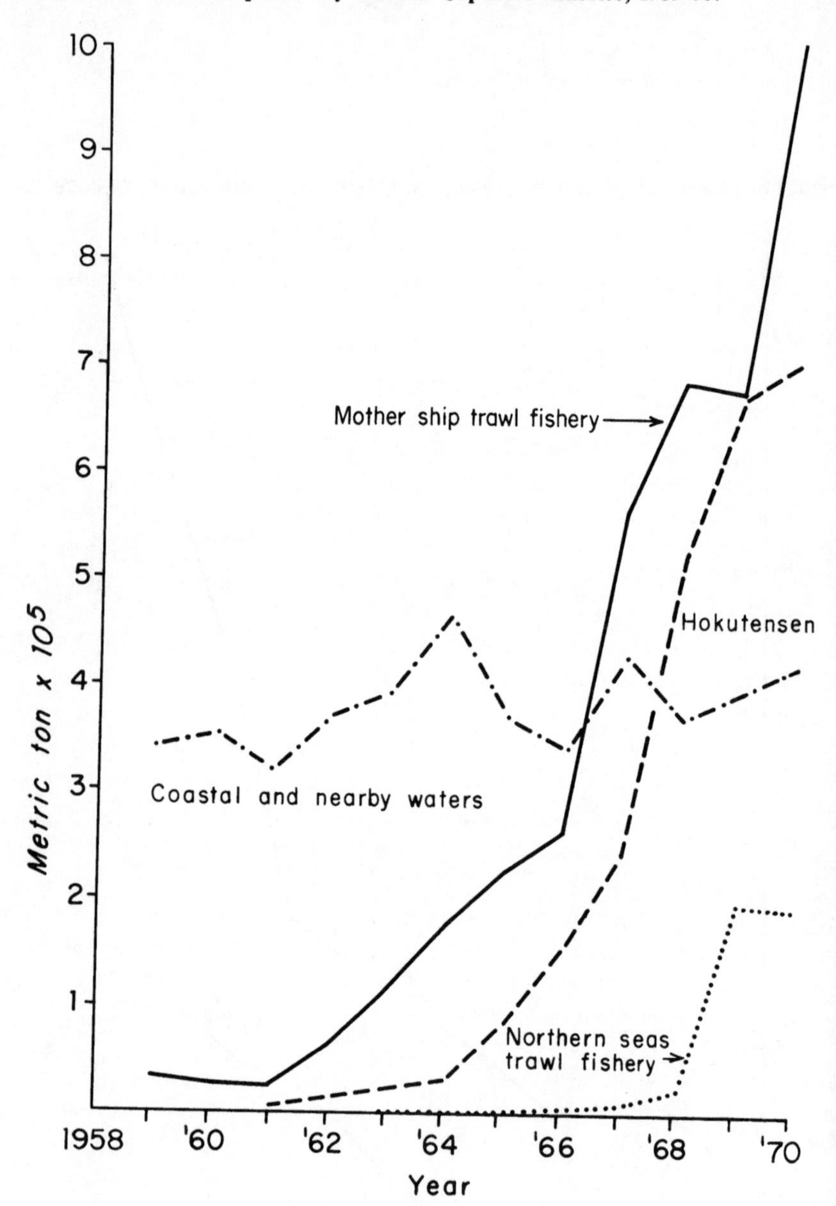

Figure 5. Catches of flounder and yellowfin sole by the Japanese fisheries from the Kamchatka-north Kurile area and the Bering Sea, 1954-70.

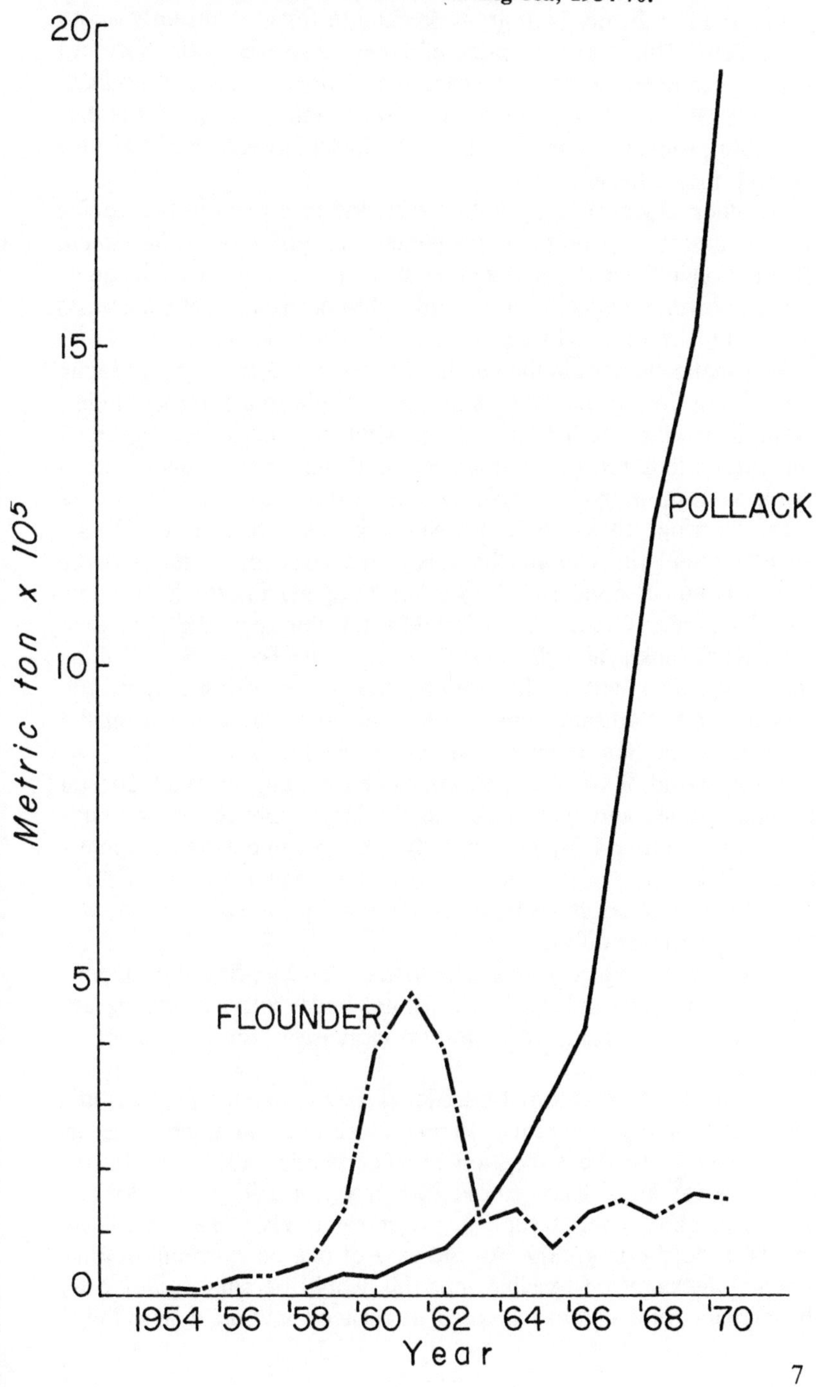

(figure 5). Pollack fishing has become very profitable because this species is used to make both frozen minced meat (surimi) and fish meal, and the former is in great demand in Japan. Currently about 85 percent of the Japanese catch of demersal species in the northern waters (Kamchatka, north Kurile, and Bering) consists of pollack. The USSR catch has also increased very rapidly; most of it is presumably processed into fish meal. A South Korean fleet has also entered this fishery.

Another aspect of exploitation of living resources in this region that is important from the management point of view is the violent fluctuation in the catches of major coastal pelagic species. The question of how to manage these resources has perplexed both scientists and administrators and solutions are yet to be found. Figure 6 shows the dramatic changes in the catches of sardines (*Sardinops*) in Japan and Korea during the 1930s and 1940s. California sardines underwent a similar decline on a somewhat smaller scale. Figure 7 indicates a long-term declining trend in the catch of spawning herring around the Island of Hokkaido since the beginning of this century. Herring stocks in South Sakhalin also went down. Many smaller populations along the Japan Sea coast of Korea and the Soviet Union disappeared. Important Asian herring stocks are now found in areas further north, namely the northern Okhotsk Sea, west Kamchatka, and the USSR coast of the Bering Sea. Taking the region as a whole, the present total catch of herring, a little less than 600 thousand tons in 1969, including the catch from the eastern Bering Sea, is much less than it used to be.

After World War II, Japan intensified fishing on other coastal pelagic species and they make up the largest portion of landings from waters around Japan (figure 8). Changes in catches of important pelagic species utilized by Japan are shown in figures 9-11. All of these stocks have been fished heavily in recent years, but their catch trends differ.

Such high-value species as salmon and crabs are fished by Japan, the Soviet Union, and the United States under international agreements but, here again, no long-term increase in the total catch is expected.

In short, the total catch of traditional species exploitable by traditional methods is apparently coming close to a maximum level in this region, with the exception of uncertainties about the future increases in the pollack catch. Not many unutilized stocks are known to exist, except such forms as sand eels (*Ammodytes*) or smelts in northern waters. An increase of one or two million tons a year might still be feasible, but this would be small relative to the present total catch estimated at thirteen million metric tons.

Figure 6. Decline of sardine catches *(Sardinops)* by Japan and Korea, 1905-58.

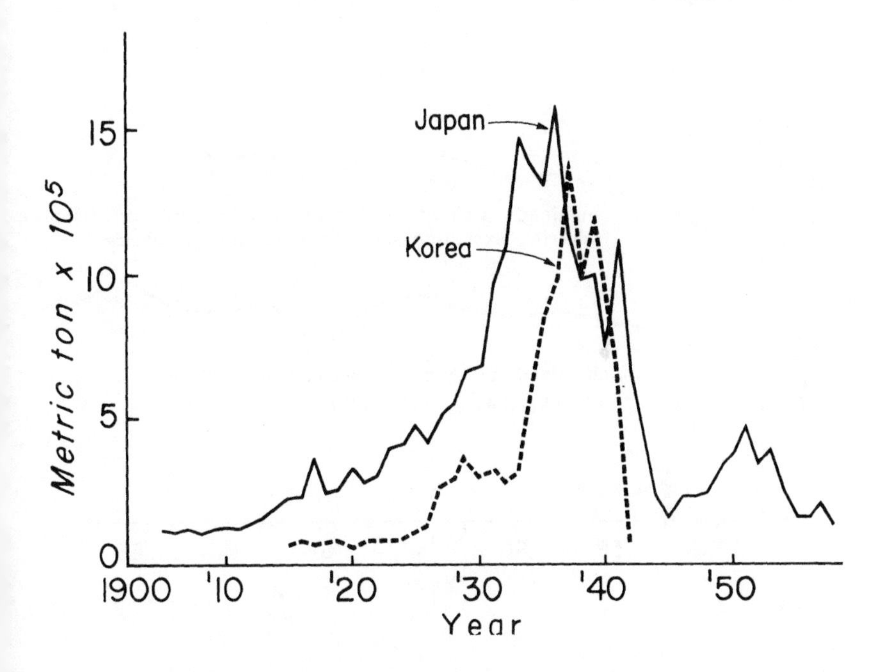

Figure 7. Long-term changes in the catch of Hokkaido spawning herring, 1871-1958.

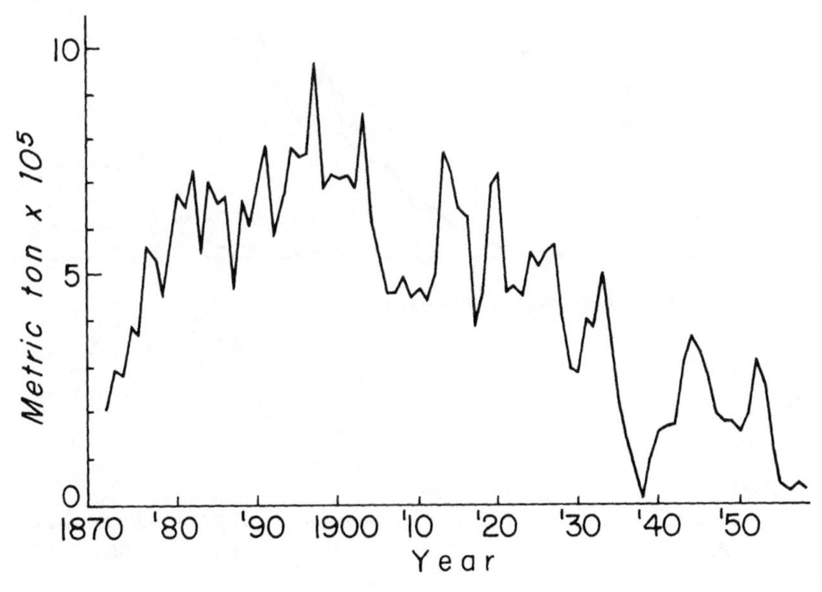

Figure 8. Combined catches of mackerel, jack mackerel, anchovy, saury, and squids by Japan, 1956-69.

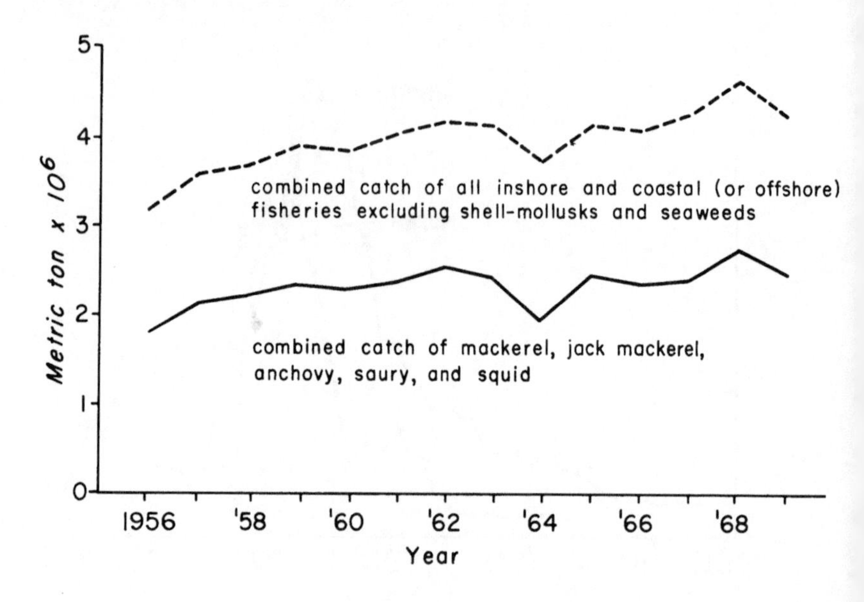

Figure 9. Catches of saury by Japan and jack mackerel by Japan, the USSR, and South Korea, 1938-69.

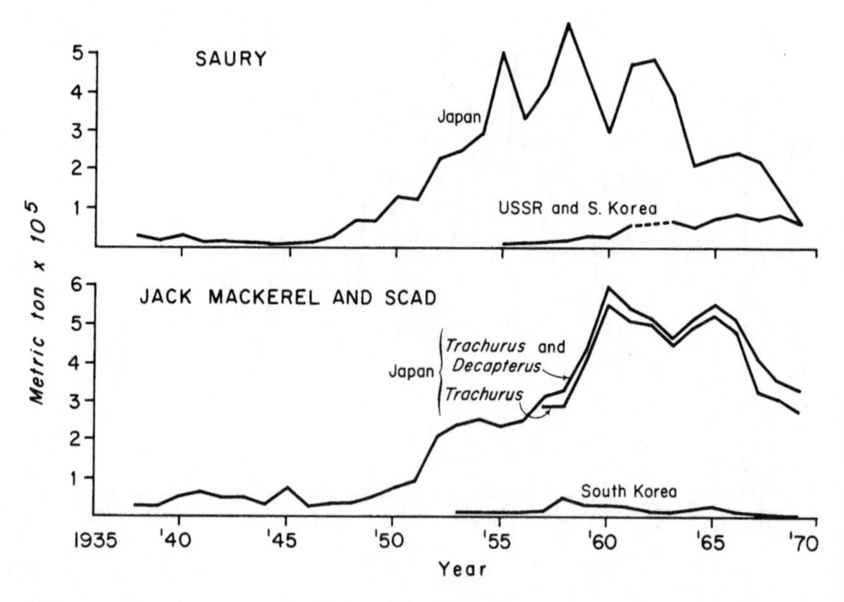

Figure 10. Catches of mackerel by Japan, and anchovy by Japan and South Korea, 1938-69.

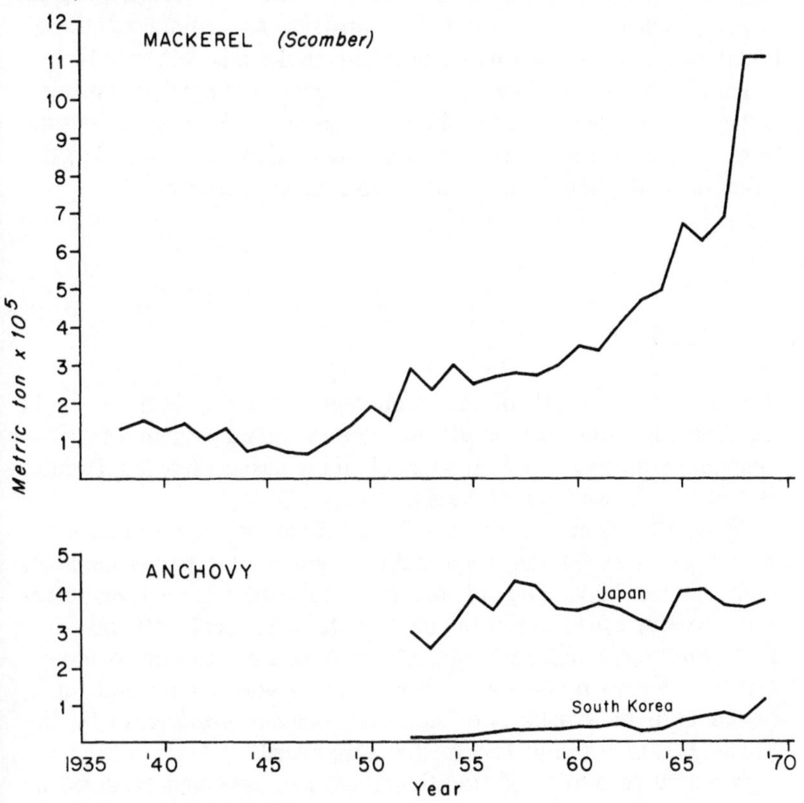

Figure 11. Catches of squids by Japan, 1938-69.

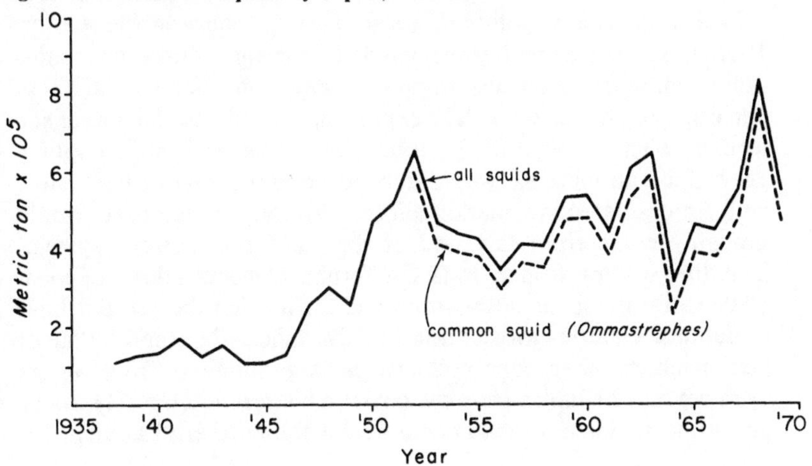

The introduction of better management measures, particularly for heavily exploited populations, might result in a measurable increase, but ever-mounting fishing pressure and complex international political and diplomatic problems make this rather unlikely, at least in the immediate future. The fisheries for pelagic species will continue to be subject to short-term as well as long-term fluctuations, because the causes of the fluctuations are still largely unknown and, therefore, are likely to remain uncontrolled.

Eastern Pacific

Waters from south of the Aleutians to Baja California are in general only moderately exploited. The estimated present total yield (including the entire U.S. tuna catch in the eastern tropical Pacific) is a little over one million tons a year.

Fishing by Canada and the United States is very selective for such species as salmon, tuna, halibut, and some other groundfish, crabs, and shrimps. The only distant-water fishery is the tuna purse-seine fishery operating in the eastern tropical Pacific off the coast from southern California to Chile. Substantial amounts of ocean perch (*Sebastes alutus*) are taken by the Soviet Union and Japan in the Gulf of Alaska and hake (*Merluccious productus*) by the Soviet Union in the waters further southward.

Potential resources of traditional species, now underutilized or completely untouched, include at least anchovy (*Engraulis mordax)* and some other species of Clupeiformes, jack mackerel (Trachurus), sole capelin *(Mallotus villosus),* saury in the western Pacific, sand eels, and some pandalid shrimps. Hake and some other groundfish may also support a moderate increase. If these and other resources were fully exploited, it would not be unrealistic to expect an increase in the total yield of several million tons a year. Such an increase would depend upon the cost of harvesting and factors affecting marketability. The largest increase would probably come from the area of the California current system, including waters around Baja California. Concentrations of most of these forms occur both within and far outside the present limit of national fisheries jurisdiction—twelve miles. Eggs and larvae of jack mackerel have been collected in large numbers from waters even beyond 200 miles from the coast (Ahlstrom 1969, p. 71); saury and some squids appear to occur across the northern Pacific.

Present Distribution of Catches

Approximate amounts of marine species, excluding whales, taken in 1969 by the different countries bordering the northern North Pacific are shown below.

Country	*Catch in million metric tons*
Japan	8.0
USSR	2.0
North Korea	0.7[a]
South Korea	0.8
People's Republic of China	1.5[a]
Taiwan	0.4
United States of America	0.6
Canada	0.1
Mexico	0.2
Total	14.3[a]

[a]Rough estimate.

Estimates are also available for the USSR and Japanese catches from the eastern half of the Bering Sea and the northern part of the North Pacific, more or less east of longitude 180°. The total combined catch is at present about 2.2 million tons (excluding Japanese tuna catches), the bulk of which comes from the shelf and slope areas of the Bering Sea.

STATE INTERESTS IN FISHERIES

As a basis for considering a future regional management regime, we will briefly review the interests of the nations that are likely to play major roles in such a regime, and indicate the relative importance of some of the fisheries and their present international implications. We will also examine some features of the domestic regulatory systems of these nations that may bear on the types of international control to be considered.

Japan

Japan had a great many inshore fisheries with a large population of fishermen even before the introduction of modern technology in the late nineteenth century. With the introduction of modern technology, more industrialized fisheries were developed through active investment. These, along with traditional fisheries, took advantage of the large domestic demand and the rich potential resources of the ocean. By the 1930s, Japan was a major fishing nation of the world, her fishermen operating not only in the waters adjacent to the home islands but also in the China Seas, Kamchatka, the Bering Sea, the Antarctic, and scattered areas in the rest of the world ocean. The industry was half destroyed during World War II when most of the larger fishing vessels were sunk by American submarines. Large fisheries developed by the Japanese in Korea, Taiwan, South Sakhalin, and the Kurile islands were completely lost as these areas became independent or were taken by the Soviet Union. Long-established historical rights to fish in Kamchatka and other Far Eastern areas of the Soviet Union were also lost. The total catch decreased to less than two million metric tons in 1945, the last year of the war.

By 1946 the industry had started to rebuild. Pressed by immediate needs to feed the nation and to earn foreign exchange, the government provided tremendous incentives for the industry to expand as fast as possible. While fishing in the nearby waters was greatly intensified, the industry also inaugurated mother-ship type fisheries in the northern North Pacific and the Bering Sea for salmon, groundfish, and king crab; a tuna longline fishery to cover the tuna grounds of almost the entire world ocean; and trawl fisheries to exploit groundfish in waters off West Africa, New Zealand, and elsewhere.

The catches of the Japanese fisheries totaled, in 1969, over 8.6 million metric tons, plus 17,400 whales with an estimated landing value of $2,419 million (at the old exchange rate of 360 yen per dollar). Marine landings consist of several hundred different species, including fishes, shellfishes, molluscs, other invertebrates, marine mammals, and various seaweeds. Each of some seventy to eighty species normally produces an annual catch exceeding 10,000 metric tons.

The economic structure of the industry is also one of extreme complexity. In 1969, there were still 66,000 nonpowered fishing boats in inshore waters. In the same year, Japan operated eleven salmon mother ships, giant trawlers (some of them over 5,000 GT), hundreds of large tuna longliners scattered all over the world ocean, and the world's largest whaling fleet. Between these two extremes were vast numbers of vessels of all kinds and sizes, which collectively contributed the largest portion of the total production.

This long-established industry, however, is now in difficulty. While maintaining high efficiency in fishing and processing, the industry has suffered from rapidly rising labor costs. Although fishermen are now much better paid than before, jobs in fisheries are no longer considered attractive in this highly industrialized country. On the international scene, too, the Japanese industry has had setbacks because of increasing unilateral claims by coastal states for broader zones of national jurisdiction, and also because of competition with other distant-water fishing nations. Readily exploitable new resources on the high seas are becoming increasingly difficult to find. Domestic demands remain very strong both for high-priced products and such undifferentiated protein products as fish meal and surimi, and the supply shortage becomes more serious every year, resulting in a phenomenal price increase (fishery products were major contributors to the high rate of inflation observed in 1971).

Alternative ways to meet the problem are examined by Kasahara (1972, pp. 277-82) in some detail. Increased imports, combined with more active investment in foreign fishing ventures, appear to offer a partial solution to the supply problem. This traditionally fish-exporting nation is becoming a major fish importer (figure 12) and this trend will continue, particularly since the nation now has a large foreign exchange reserve, some $15 billion.

In the context of international fishery problems, Japan is properly identified as a typical distant-water fishing nation, although landings are still larger from waters near her home islands (figure 13). The relative values of Japanese long-range fisheries in the North Pacific are of particular interest for the present discussion: in terms of gross values of landings, which include values added aboard ships, especially factory ships and mother ships, these fisheries account for roughly one-fourth of the total of all marine fisheries.

Two important factors affect future international regulations as far as Japan is concerned: (1) the capacity of the various sectors of the industry to adjust to or to accommodate changes in international regulations, and (2) the effectiveness of the domestic regulatory system to implement new regulations. Japan's first choice for the form of international regulations has always been free access. When conservation measures must be implemented, the Japanese preference is for free competition under limitations applicable to all parties. On the other hand, the record of fishery disputes involving Japan clearly discloses a flexible attitude toward various types of allocation.

The high-seas salmon fisheries are perhaps the most difficult to handle when introducing further changes in international regulation, particularly when these involve drastic reductions of fishing in order to meet conservation requirements or international problems con-

Figure 12. Japanese exports and imports of fishery products (excluding pearls), 1959-69.

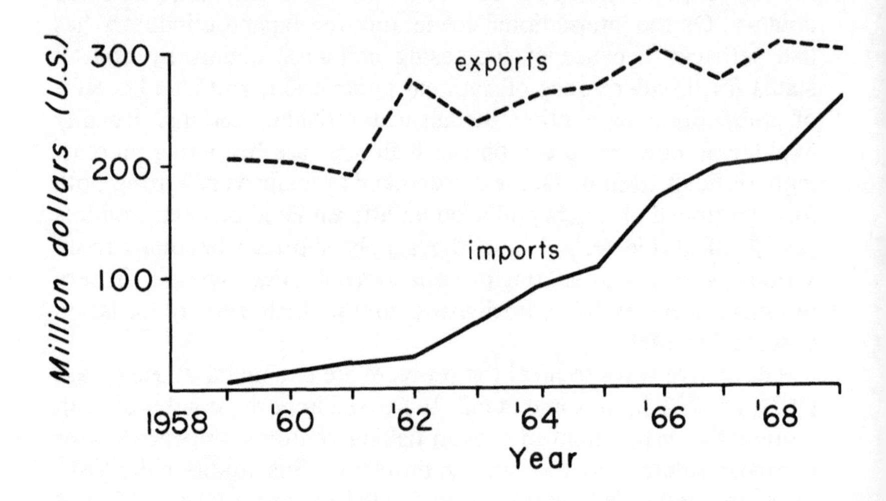

Figure 13. Production of four sectors of the Japanese fishing industry, 1959-69.

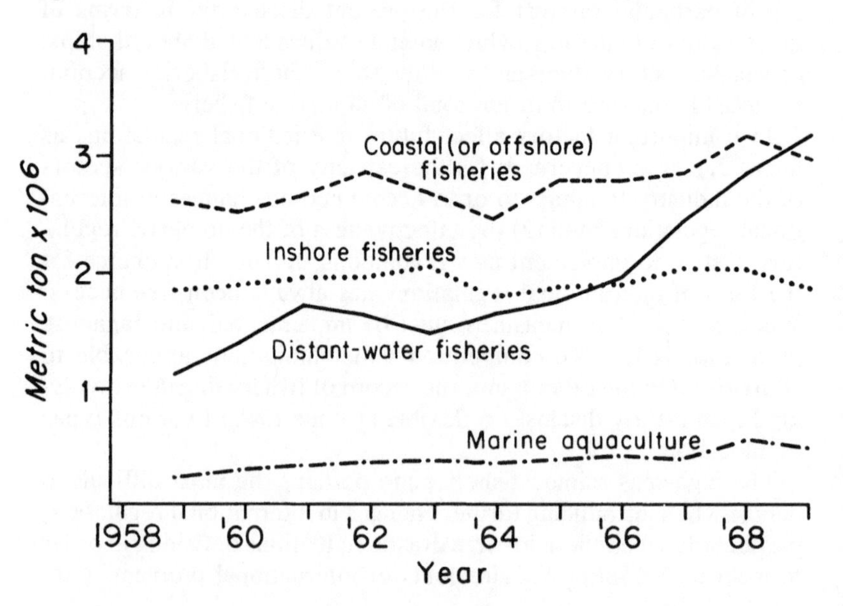

cerning allocation in the North Pacific. There are still a large number of driftnetters attached to mother ships or operating from shore bases, and also numerous small longliners. While salmon mother ships are owned by large companies, their catchers, as well as land-based vessels, belong to small companies and individual vessel owners.

The regulatory problems of the three major types of trawl fisheries are likely to differ. The so-called "hokutensen" consists of medium-size stern-trawlers, operating from shore bases. The mother-ship trawl fishery in the Bering Sea employs three types of catcher boats: Danish seiners, pair-trawlers, and otter-trawlers. The independent trawl fishery operating in the Bering Sea and the northeast Pacific consists mainly of large stern-trawlers with processing facilities aboard. The mother-ship fishery and "hokutensen" would pose more administrative difficulties in the adoption of regulatory measures than would the independent trawl fishery, for the former consists of a larger number of relatively small trawlers, most of which belong to small companies and individual owners.

The crab mother-ship fisheries would have less trouble in adjusting. Most of them are owned by a few large fishing companies, their total landing value is much less, and systems of allocation have been in practice for some time under agreements with the Soviet Union and the United States. Herring fishing off the USSR coast is now subject to strict international regulations, but the Japanese herring fisheries are not of great economic importance. Whaling is a declining industry, though its profitability appears high at the present level of catch and effort, and problems of North Pacific whaling are being handled in the context of the worldwide whaling convention. Although the Japanese and Soviet fisheries for saury and mackerel are, to some extent, competing with each other, both government and industry would like to keep them free from international regulations for the time being. Competition with South Korean fishermen has not presented a major problem in coastal waters, although some coastal fishermen have urged the government to take action—for example, the application to the northern Japanese coast of a 12-mile exclusive fishery zone under the Japan-South Korea treaty.

In short, the high-seas salmon fisheries would have the greatest difficulties in adjusting themselves to changes in international regulation. The northern trawl fisheries, particularly "hokutensen" and the mother-ship trawl fishery, would be the next most difficult ones to handle.

The domestic regulatory system of Japanese fisheries is unique and extremely complex. Basically the system consists of two parts: so-called fishing right fisheries and license fisheries. The former is further broken down into three categories: common fishing right, fixed net right, and demarcated fishing right (for aquiculture). In the context of this paper, some features of license fisheries are important (for detail see Kasahara 1972, pp. 227-82). This part of the system has been in effect almost since the inception of modern fisheries in Japan at the beginning of the century. Changes have been made toward the centralization of licensing authority (from prefectural governments to central government) and broader coverage. Virtually all offshore and distant-water fisheries are rigidly controlled by the licensing system, which regulates the activities of each fishery through restrictions on the total number of licenses to be issued, size of vessels to be used, area of fishing, method of fishing, and often species to be taken.

Application of limited entry in the management of Japanese fisheries is based on diversified considerations. Justifications used by the government for limiting entry to various fisheries also vary. Among apparent objectives are: protection of inshore fisheries against offshore fishing; reduction of competition and prevention of disputes between different groups of offshore fishermen; stabilization of fishing conditions; maintenance of profitability; conservation of resources, and prevention of international disputes. The degree of success in achieving these objectives differs from case to case, but there is no question that the system serves as a powerful and convenient means of controlling each fishery and introduces changes considered desirable by the Japanese fishery administration (Kasahara 1972, p. 228). On the international scene, the system has made it possible for the Japanese government to take a pragmatic attitude toward the question of allocation in spite of its rigid official position favoring freedom of fishing and competition under equal conditions.

Two other features of the Japanese fishing industry are worth mentioning in the same context as further means for controlling the various segments of the industry. The first is that very large companies are involved in most of the important distant-water fisheries, resulting in a concentration of ownership in many of these fisheries (Kasahara 1972, pp. 263-64). Thus, the government can handle industry aspects of most international problems concerning such fisheries by influencing these and a few other companies, some of which are subsidiaries of the large ones. The government has sometimes forced them to conduct joint operations.

The second feature is industry representation through some of the large trade associations. These include, among others: an association that represents the interests of large fishing companies engaged in distant-water fisheries and is often involved in international fishery negotiations; a federation of fishery cooperatives providing nationwide representation for Japanese fishery cooperatives; a federation of skipjack/tuna fishery cooperatives that, together with a sister association representing operators not eligible to cooperative membership, speaks for the majority of tuna vessel owners and is involved in most of the international negotiations concerning tuna fisheries; a federation of salmon fishery cooperatives representing the owners of salmon catcher boats that accompany mother ships, as well as similar associations for other salmon fisheries such as the land-based driftnet fishery and the longline fishery; and three associations, each concerned with a different type of trawl fishing in offshore and distant waters. Some of these associations have acted as signatories to nongovernmental international agreements.

Union of Soviet Socialist Republics

Before the 1917 Revolution Russian fisheries remained underdeveloped compared with some other European nations, particularly England. The bulk of the Russian catch came from the Caspian Sea and other inland waters. Seventy percent of the catch in the Barents Sea was made by British vessels and 85 percent of the salmon catch in the Russian Far East by Japanese fishermen. The total Russian catch in 1913 was estimated at only slightly above one million metric tons (Sakiura et al., 1964, p. 2).

Fisheries were largely destroyed during World War I and the period of disturbance after the Revolution, and the total catch fell to 260,000 tons in 1920. It took a number of years for the USSR government to rebuild and expand the fishing industry. By 1929 the catch recovered to the prewar level and between 1936 and 1939 it increased to about 1.5 million tons. Although the development of sea fisheries was emphasized and shore facilities were expanded, most vessels still fished in waters close to land.

The fisheries, particularly those in European Russia, again suffered seriously during and just after World War II. The systematic expansion of USSR high-seas fisheries began in the late 1940s and quickened its pace in the 1960s. Large fishing vessels were constructed not only in the Soviet Union but also in other European nations and in Japan. They were built according to a few standardized models, such as SRT trawlers, BMRT factory ship trawlers,

RTM Tropik trawlers, gigantic factory base-ships of different classes, etc. The emphasis of the postwar development has been on high-seas fisheries, largely off the coasts of other nations, with vessels operating in fleets, one fleet often consisting of more than one hundred vessels.

Expansion of fishing grounds, too, has been very systematic. After fishing in the northeast Atlantic and Barents Sea for some years, USSR trawlers appeared in the northwest Atlantic around 1954. Within a few years their fishing vessels were seen throughout the region from Labrador to the Atlantic coast of the United States. They appeared in West Africa in the late 1950s, and they have since fished intensively along the northwest and southwest coasts of Africa. In the mid-1960s USSR trawlers were also fishing off Argentina. In the Pacific, however, USSR fishing activities largely remained in waters off their own coast until 1959, when a trawl fleet entered the eastern Bering Sea. Fishing soon expanded to the Gulf of Alaska, and off the coasts of British Columbia, Washington, Oregon, and California. A few vessels have gone to waters off Mexico and South America. Boats from the Soviet Union have also fished in several other areas of the world, including the Indian Ocean. The Soviet Union is one of the two remaining major whaling nations, both in the Antarctic and in the North Pacific.

Aside from minor coastal operations along the entire Pacific coast of the Soviet Union, the Soviets conduct four major types of fishing in the North Pacific: inshore salmon fishing; trawl fishing largely off Kamchatka and in the Bering Sea; herring fishing along the Siberian coast (including Kamchatka) and in the eastern Bering Sea; and pelagic fishing for mackerel and saury.

The most valuable of all is perhaps the salmon fishery, although the catch is now at a much lower level than before the expansion of Japanese high-seas salmon fishing after World War II. The importance of salmon to the Soviet Union, Japan, the United States, and Canada will continue to be perhaps the most difficult problem for the international fishery regime of the North Pacific.

On a global basis, the Soviet Union has a larger proportion of catches in distant-water fisheries than does Japan. In the North Pacific, the Soviet Union operates distant-water fisheries in the Bering Sea for groundfish, herring, and crab, and in the Gulf of Alaska and waters further south for groundfish. On the other hand, the Soviet Union finds itself in the position of a coastal state in respect to fishing by Japan for salmon, crab, herring, and groundfish along the coasts of Kamchatka, the Kuriles, Siberia, and Sakhalin. Soviet interests in fishing in waters near the Japanese islands are at present limited to saury fishing off the southern Kuriles and northern Japan, and mackerel fishing off northern Japan, but these are relatively minor operations.

The organizational structure of USSR fisheries is briefly described by Mathisen and Bevan (1968, p. 56). All fisheries in the Pacific Ocean are under the administration of *Dalryba,* one of the five central administrative bodies in the Ministry of Fisheries. No systematic information is available concerning movements over the Pacific of the Far Eastern USSR fishing fleet, which is largely based in Vladivostok. As the Soviet Union develops deep-water trawl fishing further in the Pacific and Indian Oceans, the distribution of vessels based in the Far East might change substantially. In any case, the government is virtually in complete control of the high-seas fisheries.

Practically all fisheries of the Soviet Union are based on domestic consumption. Exceptions are such traditional export items as canned king crab, canned salmon and caviars, appreciable amounts of Pacific herring and pollack sent to Japan, and various fish landed by their vessels in West Africa. The rapid increase of fishery production, from 2.3 million metric tons in 1964 to roughly 7 million metric tons in 1970, reflects the growth of domestic consumption. There still appears to be a need for greater production of both food fish and fish meal.

Future USSR fishing activity in the North Pacific cannot be predicted with assurance. The more likely lines of development appear to be further increase in pollack fishing, entry into pelagic fishing in the California current system, and the development of living resources found on the lower half of the continental slope.

United States of America

The United States still stands relatively high in the ranking of catches by countries, sixth in 1970, but fish production by U.S. flag vessels has not increased in the last twenty years, the total catch of 2.7 million tons in 1970 being approximately the same as in 1950. This contrasts with the rapid growth of fisheries in the rest of the world during the same period. The United States, on the other hand, has always been one of the greatest consumers of fish products, and total consumption continues to increase steadily. The processing and marketing sectors of the seafood industry, being part of the much larger U.S. food industry, are far stronger than the catching sector. Because the domestic fish catching industry is unable to meet the growing demand for raw material, these sectors depend increasingly on imports. The total quantity of fishery products imported by the United States has roughly tripled during the last twenty years and, on a live weight basis, the United States now consumes more imported fish than is produced by domestic commercial fisheries.

Although the United States has an extremely long coastline extending from the Arctic to the tropics and bordering the Bering Sea and the North Pacific, its fishermen concentrate on a relatively small number of species. The landing values of shrimps, salmon, and tuna account for roughly one-half of the total, and shellfish other than shrimps make up more than one-half of the remainder. Except for tuna purse-seining in the tropics, the main fishing grounds are generally in coastal waters, and most of the vessels used are small to medium in size.

The U.S. Pacific coast fishing industry has two main sectors: coastal fisheries in the north for salmon, crabs, shrimps, halibut, and other groundfish; and the distant-water tuna fishery in the south which operates from southern California to northern Chile and is highly competitive internationally. While the coastal sector demands more protection, the distant-water section suffers from extension of national jurisdiction by other nations.

In the north, the salmon fisheries are of utmost importance. Any change in the international regulatory regime likely to increase foreign fishing of American salmon caught by Japan in waters west of the abstention line (longitude 175° West) and the threat of new entry are matters of great concern. Other international fishery problems seem relatively minor compared with salmon issues in terms of adjustments by negotiations with other states: for instance, crab fishing by the Soviet Union and Japan in the eastern Bering Sea; fishing by the Soviet Union and Japan for ocean perch, other groundfish, and shrimps in the northeast Pacific; gear conflicts between foreign trawl fisheries and local fisheries; and the effects of foreign trawling on halibut stocks.

The interests of the United States in this region, however, go beyond these immediate issues. The bordering waters of the extremely long coastline from northern Alaska to southern California are abundant in various fishery resources, many of which are exploited by foreign nations only or remain unutilized. Because of the use by foreigners, the U.S. government is under pressure from industry, as well as from the general public, to extend national jurisdiction further in order to eliminate the immediate problem of effects of foreign fishing on local fisheries and control the exploitation by foreign nations of any resource that might be considered potentially important to the United States.

It is not likely that the United States will succeed in protecting the interests of both the distant-water fishing sector and the coastal sector under a set of principles acceptable to most other nations, and if a choice has to be made, greater emphasis will probably be placed on protection of coastal fisheries. Both government and industry are fully aware that, from the fishery point of view, the United States will be one of the main beneficiaries of the worldwide trend of extension of national jurisdiction, particularly if it becomes a widely recognized principle of international fishery management.

The domestic system of fishery regulation is characterized by divided authority and a lack of means to limit entry. It is irrelevant to the purpose of this paper to discuss the merits or shortcomings of different domestic institutions as such, nor is it necessary to speculate on how the industry of each nation might have developed under a different regulatory system. But it is important to understand these differences in the context of the means available to regulate fisheries internationally. To mention a simple example, it would not create any legal or administrative difficulty for the Soviet Union or Japan to limit the number and/or size of vessels to fish in certain areas of the high seas under an international agreement. But the United States might have considerable difficulty in this respect, mostly for historical, emotional, and domestic political reasons. In fact, most of the international fishery agreements between the Soviet Union and Japan, between Japan and South Korea, or between Japan and China, limit the number of vessels permitted to operate in specified regulatory areas[1],while only one of the international agreements to which the United States is a party applies this concept to U.S. fisheries and this treaty (with Brazil) is very recent.

Industry is largely represented through various associations, most of which are relatively small but nonetheless politically influential. Vertical integration is very weak and almost nonexistent in some fisheries. Thus, the interests of the processing marketing sectors may be quite different from those of the catching sectors. It should also be kept in mind that, in Alaska and some other local areas, fisheries still provide a main or alternative source of employment; any changes in domestic institutions tend to have important social as well as economic implications.

[1]This is done either as part of the international agreements or under domestic regulations to implement them.

Canada

The mainstay of the Canadian fishing industry in British Columbia is definitely salmon, fished largely in territorial waters (i.e., territorial sea and internal waters) except for trolling on the high seas. Fisheries for halibut, herring, and groundfish are of much less importance. The development of herring fisheries by other nations beyond the present limit of Canadian national jurisdiction is only a remote possibility, for most of the herring populations are found within the limit throughout the year, with the exception of an area off the southern part of Vancouver Island. For British Columbia, the present international regime appears to be fairly satisfactory, since it protects most of the local fisheries against fishing by Japan and the Soviet Union. The main object of further extension of national jurisdiction would probably be to provide greater protection for Canada's Atlantic fisheries against foreign fishing.

There are two major differences between the domestic regulatory systems in Canada and the United States. First, fisheries regulation in Canada is mostly in the hands of the central government; in the United States authority lies with the individual component states. Second, Canada recently introduced a limited entry system to British Columbia salmon fisheries, which is restricted in scope but still more significant than any such measure in the United States. Industry interests are represented largely by a coastwide fishermen's union (which also organizes the majority of workers in shore plants) and a processors' association. They have conflicting interests, and both have a strong voice in international affairs.

South Korea

Korea (North and South combined) was a Japanese dependency from 1910 to 1945. Substantial fisheries were established with Japanese capital and technology. The sardine purse-seine fishery along the northeast coast of Korea landed over one million metric tons a year during the late 1930s, but the catch dropped to nil by 1943, when sardines almost completely disappeared from the east coast. Productive purse-seine fishing for mackerel and jack mackerel occurred along the east and south coasts. The territorial administration protected small coastal fisheries against trawl fisheries from Japan and other Japanese territories by establishing large closed areas on the west and south coasts of the peninsula.

The end of the war, however, left South Korea with a large number of antiquated small to medium sized vessels and with poor shore facilities. Devastating destruction during the Korean War further delayed development. Negotiations for normalizing relations between the Republic of Korea and Japan lasted from 1952 to 1965, a period of thirteen years. The seizure of Japanese fishing vessels by Korean authorities for alleged violation of waters under Korean jurisdiction started as early as 1949, shortly after the postwar expansion of Japanese fisheries began. In January 1952, the Republic of Korea issued a unilateral declaration claiming sovereign rights to all natural resources over a vast area delimited by the so-called Rhee Line, and enforced it strictly. The resulting controversies between Japan and Korea made fishing issues the most difficult to resolve in the lengthy negotiations.

The development of modern fisheries in South Korea began with her participation in tuna longline fishing. Initially, American capital played an important role in developing this fishery, which became a major source of badly needed foreign exchange. The Korean tuna industry has grown rapidly in the last ten years to cover much of the world tuna grounds in competition with the tuna fisheries of Japan and Taiwan.

The final settlement with Japan included large financial compensation, and substantial investments have since been made for fishery development, mainly in the form of vessel and other equipment imported from Japan under grants and loans. Distant-water fisheries are being expanded further, and coastal fisheries and shore facilities are being modernized, but it is difficult to assess the position of South Korea in the future of international fisheries in general, and those in the North Pacific in particular. There is not much room for expansion in coastal waters since resources are limited, although fisheries can be modernized and become less labor intensive. Except for laver, the seemingly great aquiculture potentials are still to be developed, but such developments would not have effects on international fishery management.

The nation is being industrialized very rapidly, resulting in greater domestic markets for fishery products and higher costs of fishing operations. Assuming that the present population is about 34 million, and that the per capita consumption of fishery products can be increased to 40 kg per annum on a live weight basis, which is not an unrealistic assumption, the South Korean domestic market could absorb, without any population increase, 1.4 million metric tons, nearly twice the amount presently consumed. Fisheries are no longer very important as a source of foreign exchange. South Korea will maintain its strength in longline tuna fishing; her share might perhaps increase further at the expense of the Japanese catch. The South Koreans are also exploring the possibility of developing a pole-and-line skipjack fishery, and they might be able to establish such a fishery in the western Pacific in competition with the Japanese.

By and large, the most promising line of development appears to be the expansion of trawl fishing, particularly in the Bering Sea and Kamchatka-Kurile waters. Such a development will certainly increase the magnitude of potential international problems concerning pollack. The extent of domestic demand for pollack and other groundfish is uncertain. Before World War II, Korea (North and South combined) annually consumed about 250 thousand metric tons of pollack, mainly in the form of processed dry fish. Today, South Korea alone might absorb this amount, whereas in 1969 the catch was less than 30,000 tons. Beyond the level of 250 thousand metric tons, growth is likely to depend on the feasibility of exporting pollack either as raw material or processed products.

The entry of South Korea into salmon fishing has caused tremendous repercussions in the United States and Japan. Although the Korean salmon venture has been discontinued, the possibility of Korea's reentry will remain as a serious potential problem that might upset the existing arrangements for salmon.

The system of fishery regulation in South Korea was originally developed by the territorial administration of Japan and was not too different from the Japanese system. The basic concept of regulation is still the same, but many changes have been introduced recently to expand distant-water operations and modernize coastal fisheries and marketing facilities. One of the deviations from the old system is direct intervention by the government in fishing and marketing through government-supported corporations, including the Korea Marine Industry Development Corporation (KMIDC) and the newly established Agriculture and Fishery Development Corporation (AFDC). Inshore and coastal fisheries are organized into cooperatives, which in turn form a national federation. Most of the distant-water operations are carried by individual vessel owners, companies, and KMIDC. (The operation of KMIDC has not been successful and the private sector is taking over its fleet.) There are no companies comparable to the biggest ones in Japan, but some of the Korean companies are now large enough to engage in operations involving factory trawlers. In short, although there are many uncertainties about the future of Korean distant-water fisheries, they are likely to become a very significant factor in future international arrangements for fisheries in the North Pacific.

Other Nations

There is no realistic basis for discussing the national interest and various aspects of the fishing industries of the People's Republic of China and North Korea. The People's Republic of China has not made any serious attempt to engage in major distant-water fisheries outside of the China Sea region. The emphasis of state enterprise in fisheries is far less than in the Soviet Union. Chinese inland fisheries and aquiculture are the largest in the world. Coastal fisheries, inland fisheries, and aquiculture appear to be forms of aquatic production suitable for the nationwide effort to provide local communities with a large degree of self-sufficiency for food. With respect to the future international fishery regime, China will continue to occupy the position of a coastal state, and will favor a greater degree of control by coastal states as a principle of international fishery regulation, preferably in the form of a wider territorial sea. A Chinese fishery association has been negotiating with a Japanese association for nongovernmental fishery agreements for East China Sea fisheries. Nothing can be said about North Korea, but there are indications that the country has plans to develop Soviet-type fisheries.

SUMMARY

In the northern part of the North Pacific, the interest of the United States and Canada will continue to lie mainly in protecting their coastal fisheries, while Japan and South Korea will maintain major interests in distant-water fisheries. The Soviet distant-water fisheries will remain strong, but the country will also be in the position of a coastal state, particularly with reference to fishing by Japan in waters off the Soviet coast. The People's Republic of China is unlikely to become a major factor in the foreseeable future except in the China Sea and adjacent areas. Salmon will continue to create more difficulties than any other resource. The large trawl fisheries conducted by Japan and the Soviet Union, with the expansion of South Korean fishing, present potentially controversial problems. At the moment, most of these fisheries are not subject to international regulation. Some of the coastal fisheries around the islands of Japan may have international problems, but compared with those mentioned above, these will be of a relatively minor nature.

II Fishery Disputes and Agreements in the North Pacific

AS INDICATED IN CHAPTER I, the fisheries of the North Pacific have long been of intense interest to the surrounding states. This interest is also reflected in the equally long history of disputes over using, regulating, and sharing these living resources. Although various methods of dispute settlement have been employed, the major emphasis has been on immediate problem solving rather than on institution building or on elaboration of absolute principles designed for all times and places. This historical pattern continues to dominate the North Pacific approach to international management efforts to this day.

CUSTOMARY LAW RELATING TO FISHERY EXPLOITATION

Generally speaking, the states on the Pacific rim do not differ fundamentally about the customary law principles currently applicable to their relations concerning living resources. There have been minor differences over the extent of internal waters—as in the differences over the Soviet claim to Peter the Great Bay—and disputes over other claims to exclusive fishing areas have been few. The United States and Japan are now alone in this region in claiming a 3-mile territorial sea and only Japan has not yet unilaterally asserted some authority over foreign fishing outside the territorial sea. No state in the region claims a territorial sea larger than 12 miles.

Enlarged exclusive fishing areas have been claimed by Korea, the United States, and Canada. The Korean extensions were extremely large and caused considerable difficulty with Japan for some years before the two states negotiated a settlement over access to the region involved. Both the United States and Canada initially created a 9-mile fisheries zone which Canada subsequently included within a 12-mile territorial sea. Special bilateral agreements have been negotiated to modify the exclusiveness of these fishery areas.

Control over living resources of the seabed has also occasioned exclusive claims and some resulting disagreement. The 1958 Convention on the Continental Shelf, to which the United States, the Soviet Union, Canada, and Taiwan are parties, provides that the coastal state has sovereign rights for the purpose of exploiting and exploring the natural resources of the continental shelf and that natural resources include certain living resources. As a consequence of this agreement, the United States and the Soviet Union, which both claim exclusive coastal rights to crabs, have had some differences with Japan, which is not a party to the convention and has historically exploited crabs on the U.S. and USSR shelves. In each instance of dispute, the legal positions have been reserved and the differences settled by negotiation resulting in a quota for the Japanese catch.

Beyond these relatively small (except for the continental shelf) areas, the states of the North Pacific now concur in recognition that the resources of the high seas are open to all who want to harvest them except as may be agreed otherwise. The major fishing states in the North Pacific, Japan and the USSR, are still strong supporters of the traditional principle of freedom of fishing on the high seas. The United States and Canada were, traditionally, also identified as proponents of this doctrine but in very recent years have moved toward recognition or promotion of greatly enlarged coastal fishing rights and, to this extent, toward modification of freedom of fishing. It is likely that the views of the North Pacific states will in the near future diverge markedly over continued recognition that the doctrine of freedom of the seas should continue to be applied to most living marine resources. One reason for this expected divergence is that in the past the most difficult fishery problems have involved accommodating claims to resources that were generally understood to be freely available to all. The United States and Canada are willing to extend coastal control over fisheries, thus removing many stocks from free availability, while the Soviet Union and Japan wish to maintain their rights of access without a need for coastal consent. Although the North Pacific has been very productive of fish under the regime of freedom of access, it has been equally productive of disputes. Within the past decade in particular, the states in this region have engaged in virtually nonstop negotiations, with differing combinations of participants, for handling a considerable array of problems. Whether extension of coastal authority over adjacent stocks would dispel all problems is discussed in chapter IV. It is worth stressing that the need for international accord in the North Pacific has not been diminished by past extensions of national jurisdiction but has rather been emphasized in even greater degree.

LONG-TERM INTERNATIONAL AGREEMENTS

The methods employed by states in the North Pacific to resolve their disputes over fisheries in the region involve both the creation of formal structures that are expected to operate for indefinite periods and the conclusion of ad hoc short-term agreements dealing with more specific problems. The latter approach is an important supplement to the former.

Fur Seal

The first notable dispute in the North Pacific arose over pelagic exploitation of fur seals and led, after nearly a quarter century of unavailing efforts at settlement by arbitration and regulation, to a multilateral agreement (The North Pacific Fur Seal Convention) between the United Kingdom, Russia, the United States, and Japan. The most significant and enduring element in the arrangement was provision for allocating the catch and its benefits among the four states concerned. Although the treaty became ineffective after 1941, it was subsequently revised in 1957, Canada now formally replacing the United Kingdom. The treaty contained a similar allocation system and, in addition, established a commission whose functions were limited to coordination of research. The commission has practically no independent power of decision and its recommendations to the four parties require a unanimous vote.

The fur seal arrangement has probably persisted from 1911 to its most recent extension because it directly faces the problem of allocating a resource and projects a formula that appears to offer each party sufficient return to justify continuing the arrangement. Perhaps the remoteness of the area involved, the Bering Sea, is also a contributing factor, since this has meant that there have been no new entrants into pelagic sealing and, hence, no threat to the established means of sharing the resource.

Halibut

Dispute between the United States and Canada over halibut began early in the twentieth century and required many years of negotiations before an initial agreement, the Pacific Halibut Convention for the Preservation of the Halibut Fishery of the Northern Pacific Ocean and Bering Seas, was reached in 1923, when an international commission was set up to recommend further regulations. The treaty has since been revised three times and the commission has compiled a notable record of scientific success in rehabilitating the stocks of halibut. At the same time this instance of management has been severely criticized because the two states concerned circumscribed the commission's objectives and competence so narrowly that the economic consequences and implications of regulation could not be taken into account satisfactorily, leading to severe economic problems.

The most startling features of the Convention for the Preservation of the Halibut Fishery of the Northern Pacific Ocean and Bering Seas of 1953, and of the three predecessor treaties, are their brevity and simplicity and, perhaps consequently, their omissions of any mention of major organizational components of the commission's structure and operation. Perhaps no other intergovernmental fishery body better illustrates the proposition that there is often a great distance between the basic constitutional charter of an organization and how it actually organizes itself to discharge its authority and responsibility. In this instance only one brief paragraph of one article (there are four operative articles) refers to the structure of the commission itself, and there is no reference to any other organ, body, or office in the treaty. Despite this lack of detail and the neglect of seemingly important procedural matters, the commission, as is well known, has a director of investigations, a staff, and a more or less formal international advisory structure and procedure for invoking it. Furthermore, none of this is new—the director's office and staff were established immediately after the 1925 treaty came into effect and they have been exceedingly important in the commission's functioning ever since.

The Halibut Commission's given objective is maximum sustainable yield (MSY). This, of course, is not unusual since practically all international fishery conservation agreements postulate the same, or a nearly identical, goal. However, the commission's performance in seeking its goal has been especially controversial and the selection of the MSY objective has been severely criticized in studies that have examined and documented the harmful economic impact on the fishing industry of regulations aimed primarily at a biological objective. Although these strictures originally concerned only the commission's regulatory system, they have since been generalized to many other fisheries and are now very widely regarded as cogent diagnoses of a management system that has pursued goals too narrowly conceived.

In sum, the Pacific Halibut Convention seems more significant, at least in organizational terms, for what it does not say than for what it does. The decisions and practices of the commission are far more important in this respect than the provisions of the treaty. However, with respect to the commission's regulatory objective and measures, the treaty still seems to be authoritative and reflects the limited expectations of the parties to it.

Salmon and other species

Canada and the United States. The earliest dispute in the eastern Pacific, that between Canada and the United States over Fraser River salmon, arose about the turn of the century and culminated in 1937 with the Convention for the Protection, Preservation and Extension of the Sockeye Salmon Fisheries of the Fraser River System, which created one of the pioneer international fishery management bodies, the International Pacific Salmon Fisheries Commission.

The salmon convention closely resembled the halibut convention in its relative lack of detail on the structure and operation of the commission it established. The only article in the convention that dealt with the structure of the commission merely established that it should have six members, three from each party, specified that they were to be appointed by the President of the United States and the Governor General in Council of Canada and to hold office at the pleasure of the appointing party, and provided that the commission should exist as long as the treaty remained in force. Article II also provided that each party should pay the salary and expenses of its commissioners and that the parties should be equally responsible for payment of joint expenses. There was no reference to any subsidiary bodies, to any other officials of the commission, or to a staff, nor were there any directives concerning commission operations, except that Article VI provided that commission action pursuant to the treaty required the affirmative vote of at least two commissioners from each party.

The commission immediately employed a director of investigations who in turn recruited a staff to carry out the inquiries enjoined by the treaty. The staff soon came to comprise a substantial body of people, as was surely necessary if the commission were to discharge the complex tasks assigned to it.

The objective of the 1937 salmon convention was to restore a fishery that had been badly damaged by human and natural forces. The preamble records "that the supply of this fish in recent years has been greatly depleted and that it is of importance in the mutual interest of both countries that this source of wealth should be restored and maintained. . . ." In practice it is understood that the commission's objective is interpreted to mean securing the maximum production of salmon, an enterprise of extreme complexity and difficulty.

In addition to the difficult environmental and biological problems, proper management of the runs is greatly complicated by the objective specified in the treaty of assuring that Canadian and U.S. fishermen share equally in the annual catch. This provision for allocating the catch equally between the two parties dealt explicitly with what is acknowledged now to be the foremost problem in international fishery management. The rationale for this unusual provision was explained in Article VII: "Inasmuch as the purpose of this Convention is to establish for the High Contracting Parties, by their joint effort and expense, a fishery that is now largely nonexistent, it is agreed by the High Contracting Parties that they should share equally in the fishery." It seems obvious that this rationale for an allocation arrangement is unique to this situation since such explicit provisions are still exceptional, although states are now beginning more and more to handle this problem directly.

In its first two decades of operation, the convention provided the commission with independent decision-making authority in limiting or prohibiting the taking of sockeye salmon within the conditions specified in the treaty. Subject to certain provisos that do not bear on this point, Article IV empowered the commission "to limit or prohibit taking sockeye salmon in respect of all or any of the waters" in the convention area and stated that such orders "shall remain in full force and effect unless and until the same be modified or set aside by the Commission." Article V provided that the commission could "prescribe the size of the meshes in all fishing gear and appliances" that could be operated in the convention area.

This unusual delegation of competence was partially rescinded in the 1957 Protocol on pink salmon, which added a paragraph to Article VI of the convention: "All regulations made by the Commission shall be subject to approval of the two Governments with the exception of orders for the adjustment of closing or opening of fishing periods and areas in any fishing season and of emergency orders required to carry out the provisions of the Convention." This appears still to leave the commission with a useful degree of independent authority since there is frequent occasion to issue the orders mentioned.

Even with this modification, the salmon commission is entrusted with a wider, combined range of decision-making and operational functions than any other comparable body. On the latter side, as already noted, the commission undertakes its own research operations and these not only embrace an enormous region of direct interest to the commission but also require the recruitment of a very considerable mixture of skills and talents. Although the commission's reports are not generous in disclosing the details on such mundane but vital subjects as staff size, composition, or even commission expenditures, the 1970 report does at least list the staff, which then numbered fifty-five. It is obvious from a survey of the annual reports that the staff is assigned a variety of highly sophisticated research duties.

The viability of the salmon convention is perhaps best indicated by the fact that since 1947, a quarter of a century ago, it has remained in effect wholly on a year-to-year basis. All told, the convention regime is now over four decades old.

Alaska. The controversy over the attempt to reserve Alaskan salmon to U.S. fishermen alone began just prior to World War II but the treaty which partially achieved this aim came after the war. In 1951 the United States, Canada, and Japan concluded an agreement establishing both the International North Pacific Fisheries Commission (INPFC) and the principle of abstention. In accord with the latter principle, Japan agreed to refrain from fishing for salmon of American origin east of a provisional line drawn at longitude 175° west. Japan also agreed to abstain from fishing for halibut and herring as well as other stocks that might be added in accordance with principles delineated in the treaty. This arrangement for salmon is the major purpose of the treaty but it has been only partially realized because American salmon migrate west of the abstention line where they are caught in measurable numbers by the Japanese high seas fleet. The parties have been unable to come to further agreement, as the treaty provides, upon moving the provisional line, but they have agreed to remove certain stocks of halibut and herring from the abstention requirement (Kasahara 1972, p. 247). No stocks have been added to this requirement.

The parties to the INPFC agreement have sharply differing views on its allocation provisions and it is noteworthy that Japan continues to abide by the treaty and to refrain from terminating it.

Western Pacific. Japan has fished stocks of salmon produced on the Asian mainland for a century, first from stations on Russian territory but since World War II only on the high seas. Numerous agreements between Japan and the Soviet Union have been instrumental in resolving disputes over this resource, the last and most important being the Convention Concerning the High Seas Fisheries of the Northwest Pacific Ocean of 1956. The primary aim of this convention, which applies to all but territorial waters of the northwest Pacific, was to provide for allocation of salmon and regulation of Japanese salmon fishing on the high seas. Although this agreement primarily concerns salmon, it also provides a mechanism in the form of a joint commission for seeking the parties' objectives in relation to all living resources in the northwest Pacific. In implementing the convention the commission is supposed to determine the catch limit for the Japanese take of salmon, to change the regulations contained in the convention, and make recommendations for research and for conservation of the fishery resources of the area. Sometimes the commission itself is bypassed and negotiations on the annual catch limit are conducted by high officials of the two governments. It is also noteworthy that the Japanese quota has steadily declined and that the commission is devoting increasing attention to other stocks in the area.

Japan and the Republic of Korea

The most recent arrangement made in the North Pacific region terminated thirteen years of negotiation between Japan and the Republic of Korea, lasting from the end of World War II until 1965. In agreeing to set up the Japan-Republic of Korea Joint Fisheries Commission, the two countries provided both for recognition of certain principles and practices in regulation and for establishment of an institution to facilitate cooperation. With respect to the former, the agreement provides for mutual recognition of the right to establish a 12-mile fishing zone over which the coastal states would have exclusive fishing jurisdiction. In addition, the parties established a joint regulatory area, beyond the exclusive zone, in which fishing by both sides is regulated as to numbers, sizes, and types of fishing vessels as well as annual catches. Other measures include closed areas and seasons in various locations.

The commission's goal is to ensure the maximum sustainable productivity through conservation and rational exploitation. Its authority extends to the joint regulatory area and also to a joint resource survey zone whose extent, and the survey activities pertaining to it, are decided on the basis of commission recommendations. The commission may also make recommendations on matters concerning provisional fishery regulations and penalties for violations of the agreement. The commission meets annually and its formal actions in making resolutions, recommendations, and decisions require the concurrence of both parties.

BILATERAL, SHORT-TERM AGREEMENTS

Nine ad hoc bilateral agreements are currently in force in the eastern Pacific north of Mexico. These deal with a considerable variety of regulatory problems including gear conflicts, access to areas subject to national jurisdiction, allocation of stocks and limits on effort, research activities and dissemination of data, and visits aboard vessels. Generally these arrangements concern specific problems over a short time period, hence they are renegotiated frequently.

There are two main western Pacific bilateral agreements, one on a governmental and the other on a nongovernmental level. Japan is a party to each of these agreements, the other parties being the Soviet Union (the Convention Concerning High Seas Fisheries of the Northwest Pacific Ocean) and China (nongovernmental agreement).

The arrangement with China grew out of the large number of seizures of Japanese trawlers by China that began with the outbreak of the Korean War. Chinese patrol vessels seized over 150 Japanese fishing vessels in the period 1951-54. In 1954 negotiations between a Japanese private group and a Chinese fishery association led to a one-year agreement on the trawl fishery in the East China Sea (including the Yellow Sea). Trawling in coastal areas was prohibited and limits were established on the numbers of Japanese and Chinese trawlers permitted to operate in certain areas. After an interval during which the agreement was terminated, it was revived in 1963 and has been revised and modified from time to time. In December 1970 the two sides agreed to regulate purse seining in certain areas, closing one area and limiting the number of seiners in two others.

The Japan-Soviet Union agreement of 1969 on crab fishing grew out of the activities of the Japan-Soviet Northwest Pacific Fisheries Commission. Partially in response to the commission's activities, the two governments had taken various measures from 1958 to 1968 to restrict fishing and to avoid conflicts, including allocation of fishing grounds between the fleets of the two nations. In 1969, after the Soviet Union proclaimed sovereign rights over the natural resources of the continental shelf, the two states began negotiations which culminated in a separate agreement on crab fishing. As in the case of the Japan-United States crab agreement, the disputed legal position is left unresolved; the parties agreed to limit the number of vessels and to establish catch quotas (in terms of either canned crab produced or the number of crabs caught) for particular species in named areas. Fishing grounds for the various crab fisheries were specified, and closed seasons, size limits, and other restrictions were provided.

CONCLUSION AND SUMMARY

The preceding brief discussion shows that fishery controversies in the North Pacific are resolved on a case-by-case basis, without regard for any general overall management scheme or institutional arrangement. There are a great many explicit agreements for resolving one or another of a variety of problems, only a few of which have all participants in a fishery as parties. The specific objectives sought by the various fishery agreements are sometimes similar and sometimes very different, but even the similarities are submerged by the considerable variations in the implementation of practical measures. For example, allocation is sometimes an explicit goal and sometimes only implicit, but the methods for dividing differ greatly. One may contrast, for example, the sharing prescribed in the Fraser River salmon and fur seal agreements with the exclusion established in the International North Pacific Fisheries Commission.

Organizational structures in the North Pacific also vary in some respects, although they are generally simple, reflecting an inadequate delegation of authority. The earliest established bodies, for halibut and salmon, are distinguished by having full-time research staffs. The point to emphasize, however, is that the creation of independent research staffs apparently was considered feasible or desirable only in limited circumstances, characterized most notably by friendly neighboring states of similar culture, language, and systems of government.

Whatever the nuances of particular situations, the most outstanding feature of international regulatory bodies in the North Pacific is the reluctance of national governments to delegate any of their authority on fisheries to an international agency. In every instance noted in this discussion, and in most instances elsewhere in the world, the ultimate authority to prescribe and to apply policy is retained by the nation-state. Even with respect to apparently less important decision functions, such as monitoring conditions of fish stocks and their exploitation, there is no discernible trend toward conferring much authority on an international body or even toward adequate coordination of national efforts. In terms of control over finances, fishery bodies are noticeably deprived, budgets being notoriously slim.

Thus, whatever the variability in the structure and functioning of the fishery bodies of the North Pacific, the record is clear that ultimately national governments have been relied upon to make critical decisions. It is perhaps going too far to emphasize a correlation between this and the inadequacy of fishery management in this region. We turn now to the remaining outstanding management problems.

III Contemporary Problems of Fishery Management in the North Pacific

AS BRIEFLY DESCRIBED IN CHAPTER II, the recorded history of controversies over fishery exploitation in the North Pacific is not quite as old as that in the North Atlantic, but still extends back almost a century. The complicated network of fisheries conventions and agreements resulting from the long succession of disputes has not resolved all the problems of the region. While the methods employed for resolving North Pacific fisheries disputes and the agreements produced thereby exhibit both shortcomings and merits, on balance the system needs alteration to cope with emerging and expected problems. Of course the advantages of the existing arrangements should be retained but changes must be introduced to alleviate major defects. The following discussion highlights the major problems that confront the present system and poses the question of what should and can be done to improve matters.

AD HOC NATURE OF ARRANGEMENTS

The previous chapters supply abundant evidence that fishery management problems have been resolved strictly on an ad hoc case-by-case basis. Without significant exception, the various North Pacific arrangements were fashioned as seemed to be necessary to resolve specific controversies rather than developed as part of a comprehensive, systematic approach to international fishery problems (real or potential). The prewar fishery agreements between Japan and Russia (later the Soviet Union) were largely made to resolve controversy over Japanese salmon fishing from Russian ports. Japanese fishing for crab and herring was also covered but this was a minor issue compared with the salmon controversy. The later 1956 Convention Concerning the High Seas Fisheries of the Northwest Pacific Ocean between Japan and the Soviet Union was an immediate result of, and response to, a unilateral Soviet declaration aimed at control of Japanese high-seas salmon fishing (see figures 14 and 15). Although, as noted, the constitutional scope of the treaty is comprehensive enough to handle any problem in the northwest Pacific as far as these two states alone are concerned, actual regulations under the treaty have been applicable only to salmon, herring, and crabs. Furthermore, crab problems were separated out from this arrangement for treatment in a new agreement in 1969.

The two fur seal treaties in 1911 and 1957 dealt with only one species of animal. While they have successfully established a unique concept of allocation, they remain extremely limited in scope.

The tripartite International Convention for the High Seas Fisheries of the North Pacific Ocean that set up the International North Pacific Fisheries Commission (INPFC) in 1951, was based on the abstention principle, and was originally negotiated as a means for the United States and Canada to eliminate the threat of Japanese fishing for salmon, herring, and halibut in the eastern half of the Pacific. A list specifying the stocks qualifying for abstention has always been the most important part of the agreement. Among the three forms included in the original list, herring are no longer an important issue, as only the stocks in southern British Columbia remain on the list. Abstention from high-seas halibut fishing only by Japan no longer makes much sense, since the Soviet Union can and does catch halibut legally in the waters from which Japan is obliged to abstain from halibut fishing. The INPFC has extended its scope somewhat by including discussion on research and stock assessment aspects of king crab, and later tanner crab, in the eastern Bering Sea. But, in this case also, management aspects of crab fishing have been taken away from the commission to be covered under a separate agreement. The commission also does a certain amount of monitoring for trawl fisheries in the Bering Sea and northeast Pacific, but has refrained from making any recommendations except on halibut. Thus, for practical purposes, the treaty continues to exist largely to maintain the status quo for salmon.

Three sets of bilateral arrangements are in effect to regulate crab fisheries in northern North Pacific areas: Those between Japan and the Soviet Union (see figure 16), Japan and the United States, and the Soviet Union and United States. Even in one small part of the area, the eastern Bering Sea, crab fisheries (king crab and tanner crab), are regulated by three agreements: in addition to the Soviet Union-United States and the Japan-United States agreements, there is a supplementary agreement between Japan and the Soviet Union (see figure 17) concerning allocations of tanglenet fishing grounds, which are within U.S. jurisdiction (the continental shelf).

The 1937 Fraser River salmon convention is designed only for

Figure 14. Salmon regulatory area for the 1956 season and initial salmon regulatory area under the 1956 Japan-Soviet Union treaty.

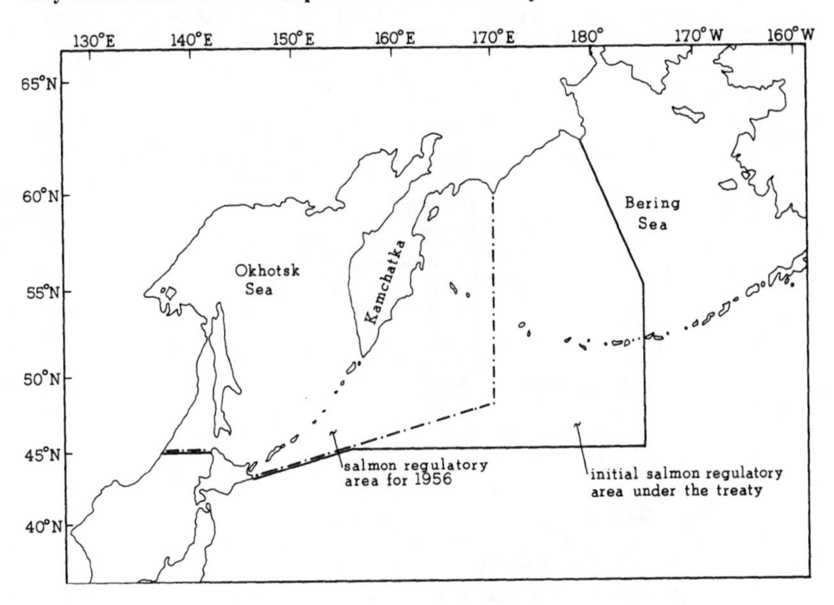

Figure 15. Allocation of high-seas salmon fishing grounds, 1969.

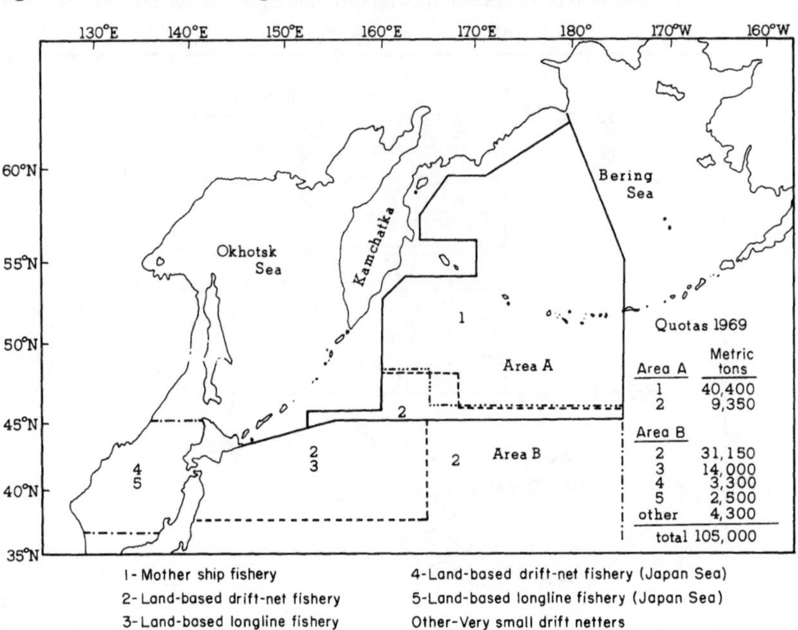

Figure 16. Crab fisheries regulated under the Japan-Soviet Union crab agreement, 1969.

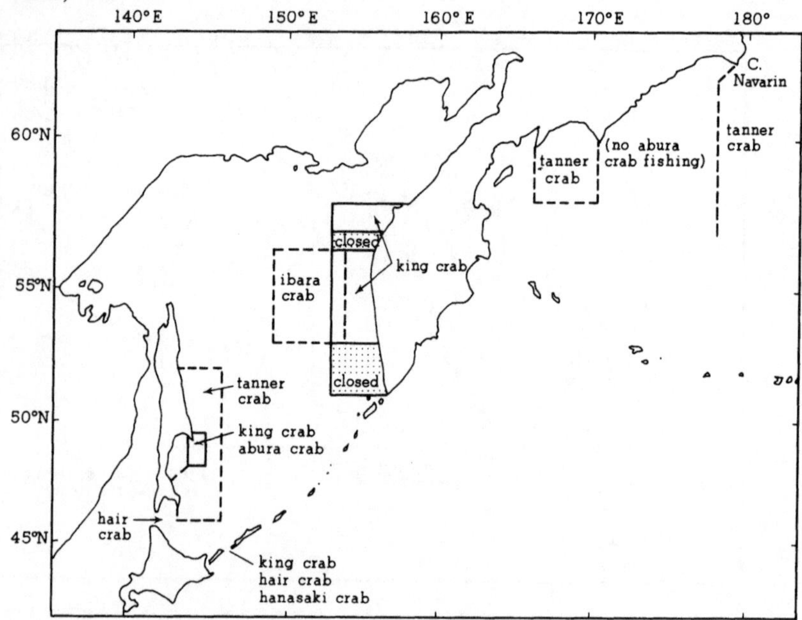

Note: **King crab fishing grounds off west Kamchatka are allocated between the mother-ship fleets of the two nations.**

Figure 17. Allocation of king crab fishing grounds in the eastern Bering Sea, 1969.

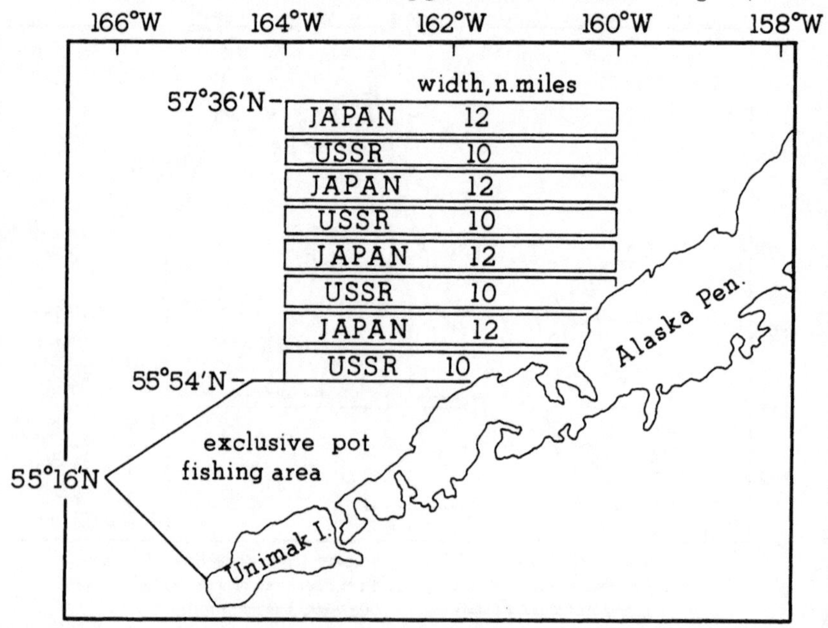

this particular salmon system and the 1923 halibut convention only for halibut along the coast of the United States and Canada. Even the Japan-South Korea agreement and the nongovernmental agreement between Japan and China are far short of covering major fishing activities in the East China Sea and the Yellow Sea as a whole. The former sprang from a unilateral jurisdictional claim by South Korea, and the latter was negotiated to reduce conflicts between Japanese and Chinese fishermen concerning trawl fishing activities, with purse-seine regulations added later.

LACK OF COMPREHENSIVENESS

While the present regime thus consists of a complex network of ad hoc specialized arrangements, some of the largest fisheries in the area that have real or potential international implications escape coverage by any of the existing agreements. Most of the trawl fisheries conducted by the Soviet Union, Japan, and South Korea in the Bering Sea, the northeast Pacific, as well as off Kamchatka and the northern Kuriles, are not subject to any international regulation (except minor restrictions under bilateral agreements arising from exclusive fishing claims by the United States and Canada). The present total catch of these fisheries is estimated to exceed three million metric tons. Other large fisheries that are *internationally* unregulated include, among others, herring fisheries conducted by the Soviet Union and Japan in the central and the eastern Bering Sea; saury and mackerel fisheries in the western Pacific conducted by Japan, the Soviet Union, and Korea; and squid fisheries conducted by Japan and South Korea.

Furthermore, much of the trawl and purse-seine fishing in the East China Sea, including the Yellow Sea, is unregulated except for activities in regulatory areas covered by the Japan-South Korea agreement and Japan-China agreement. There are no international regulations concerning tuna fisheries in any part of the Pacific except for yellowfin tuna in the eastern tropical Pacific.

The result is that, despite the numerous international treaties and agreements, more than 90 percent of the total catch in the North Pacific comes from fisheries that are outside international regulation. Although it is true that this state of affairs reflects the absence of an international issue concerning the unregulated fisheries, the most likely sources of future difficulties are not in any way anticipated by the present system of regulation. To take pollack fishing as an example, there is no mechanism to resolve the international problems likely to arise from fishing in northern waters by Japan, the Soviet Union, South Korea, and possibly the United States. Many other examples can be cited.

EXCLUSIVENESS

The shortcomings represented by the ad hoc approach to fishery problems and by the lack of comprehensiveness in these arrangements are magnified by a third characteristic: none of the fishery agreements and treaties for the North Pacific alone are open to new entrants, either formally or effectively. In practically every instance, admission of third parties would require the renegotiation of the entire agreement, since each is custom designed to fit the original parties in a very specific context. A natural outcome of this common feature of existing agreements is that the question of new entry cannot be handled within their framework; the problem must be dealt with through a new round of direct negotiations. Thus, in a region where over 90 percent of the catch comes from internationally unregulated fisheries and where only one agreement (fur seal) involves all the major states concerned, it would be virtually impossible to handle probable problems involving a new fishing entrant by recourse to an existing mechanism.

PROBLEMS OF MONITORING AND RESEARCH

The existing pattern of agreements either bind differing combinations of parties, or are specialized to one or a very limited range of species, or are confined to resolution of a narrow and specific problem. Therefore, issues are usually not negotiated until they become very serious. The patchwork of agreements does not, in sum, constitute an overall system for monitoring new developments and for calling the attention of the nations concerned to contemporary problems, biological or political, that keep arising in this region as fishing activities in international waters become increasingly intensive. Indeed, it is conceivable that the number and complexity of existing arrangements, which require unending preparation and virtually ceaseless negotiation, seriously inhibit any capacity there may be in the system or in individual nations to take an overall look at developing problems or difficulties.

The acuteness of specific problems, national or international, also influences the direction of research. While much work is done to determine the effect of trawl fishing on the small North American halibut stocks utilized by a longline fishery, little international effort is devoted to evaluation of the status of exploitation of much larger stocks, such as pollack or flounder in the eastern Bering Sea. On the other side of the Pacific Ocean, some 1.5 million tons of pollack are landed by the USSR and Japanese fleets in the Kamchatka-Kurile waters, but no attempt has been made to initiate a joint study of the status of this important resource. In general, under the present system, no serious international investigation is conducted until problems of a political nature develop and by this time the costs of resolution may be unnecessarily high.

This shortcoming is intensified because, through modern technology, a large fishery may suddenly develop. A new resource may be exploited to a maximum level within a few years —sometimes in two or three years—as has been demonstrated in the North Pacific, the northwest Atlantic, east central Atlantic, and elsewhere. The traditional concept of management based on time-consuming research is often unworkable under these new conditions; scientists have yet to find effective ways to cope with contemporary management problems arising from the rapid development of fisheries.

Another common problem is the lack of a uniform system of collecting and analyzing statistical data. Each of the countries bordering the North Pacific has a relatively well-developed national statistical system, but each is substantially different from the others. Even for the stocks commonly exploited by two or more nations in one area, there is now no way to collect uniform catch and effort data, or comparable biological information, from the countries concerned (the only exceptions are certain stocks jointly managed by the United States and Canada).

Even when there is comprehensive information on the North Pacific there is a problem in making it accessible. Most of the detailed fishing and biological data are made available only to those directly engaged in research programs required by the respective international agreements. Consequently, it is difficult for outsiders to make contributions to the solution of international management problems through research based on such data.

ALLOCATION

There is no established set of principles or rules, or even a widely accepted formula, concerning allocation as applied to the international management of living resources of the sea. On the other hand, it is becoming increasingly difficult to develop an acceptable management regime without taking into account the question of allocation (who gets what and how). For the time being, specific systems must be developed to meet specific problems.

One positive aspect common to practically all international fishery agreements in the North Pacific is that, whether or not an agreement is explicit about it, the question of allocation is dealt with intensively.. This is particularly obvious in the agreements to which Japan is a party (Kasahara 1972, p. 262). The official position of the Japanese government concerning high-seas fishing has always been to support freedom of fishing and, where conservation measures are required, free competition among nations with limitations equally applicable to all. In practice, however, Japan has accepted various forms of allocation as a means of accommodating the conflicting interests of the nations concerned, although Japan has seldom taken the initiative for making such arrangements. The most extreme existing form of resource allocation is "abstention" as provided in the 1951 North Pacific treaty establishing the INPFC, which Japan accepted, though under unusual circumstances. In most other cases, allocation has been implemented through a combination of catch quotas, sometimes with a system of allocating fishing grounds, and through direct control on fishing effort. This general feature of international fishery regulations in the North Pacific appears to be a natural outcome of the fact that the existing agreements have resulted from immediate problems of a political nature, in addition to, or sometimes rather than, common concern over the state of the resources under question. Furthermore, the domestic institutions of the Soviet Union, Japan, South Korea, and China are such that direct control on entry of their vessels into a specific fishery or area do not create insurmountable legal or political problems in their own countries. This characteristic of the North Pacific arrangements is one which might be maintained and made more explicit in future agreements.

DISTANT-WATER FISHERIES VS. COASTAL FISHERIES

The several bilateral agreements in the North Pacific that aim at accommodating the interests of distant-water and coastal fishing states have already been discussed. These are ad hoc, short-term executive agreements and although they serve as temporary measures to adjust fishery relations disturbed by extended national jurisdiction, we do not believe that this problem can be resolved in this way in the long run.

The problem should be viewed in the wider context of world trends regarding fisheries. International conflicts between fisheries of coastal and of noncoastal states are neither new nor confined to the North Pacific. In the North Pacific these appear to be less serious than the conflicts in the east central Atlantic, where fleets from about twenty African and non-African nations fish, or in the northwest Atlantic, where many European nations operate fleets off Greenland, Canada, and the United States.

But in the North Pacific, too, most of the international fishery problems after World War II have arisen from the conflicting interests of distant-water fishing states and coastal fishing states, particularly since the two greatest distant-water fishing nations in the world are engaged in intensive exploitation of resources in this region. The postwar expansion of the Japanese and USSR distant-water fleets has caused more international fishery problems in the North Pacific than any other single factor. To be fair in assessment, however, credit should also be given to these two nations for their contribution toward the development of new resources in the Pacific and all over the world. For example, the Soviet Union and Japan developed demersal resources in the northern North Pacific that now support a combined annual yield estimated at three million metric tons. These would have remained unexploited or grossly underexploited without the effort of these two nations. Japan pioneered in developing the tuna resources of the Pacific and of the world ocean exploitable by longlining. The Soviet Union and Japan initiated large-scale fisheries along the west coast of Africa and elsewhere around the world.

On the international scene, these two nations are seldom given credit for their contribution toward resource development because the impact of their distant-water fisheries on high-seas resources, some of which are also utilized by coastal states, has been such that many nations consider USSR and Japanese fishing to be major factors responsible for the depletion of resources on a global basis. Whether or not actual overfishing occurred, the way new resources have been utilized by these nations appears frightening to others. A new stock may sometimes be exploited to a maximum level in only two or three years. Effort is moved from one resource to another, or from area to area. This new pattern of fishing, characterized by concentration of effort through large fleet operations and shift of emphasis from one resource to another, is not necessarily a bad strategy from the point of view of maintaining the total production and the profitability of the specific industry using these tactics. But such a pattern is not attractive, or even acceptable, to many other nations because it contradicts the established principles of management based on the concept of maximum sustainable yield and, more important, because such a pattern can be adopted only by nations having a well-organized distant-water fishery capability (Kasahara 1972 p. 266). If a coastal nation is unable to participate substantially in the utilization of a resource for technological or economic reasons, it frequently has a strong desire to keep the resource underdeveloped rather than see some other nation exploit it.

This attitude is not simply inspired by a wish to deny to others what one cannot have, but rests on basic considerations pertaining to the success of the coastal nation entering the fishery after it has first been exploited by a distant-water fleet. By the time a coastal fishing nation is ready to utilize resources in nearby waters, their abundance may have been decreased by distant-water fleets to a point at which the operations of the coastal fishermen, in most cases not as efficient as those of the distant-water fleet, are not profitable. This does not necessarily mean that overfishing in a biological sense has occurred. In order for any stock to be exploited to a substantial degree, its abundance (and hence the catch per unit of effort) is reduced to a varying degree, depending upon the biological characteristics of the stock and the nature of the fishery exploiting it. There are exceptions, but a rather sharp decrease in catch per unit of effort should be expected from intensive exploitation of most fish stocks, particularly at the initial stages. But the coastal state's fishing of the stock may be handicapped, and therefore made less profitable, by the earlier development of a distant-water fishery based on the same stock.

In these circumstances, the desire to exclude distant-water fleets can become irresistible when the distant-water nations themselves are adopting various measures to avoid or minimize conflicts between their own coastal fisheries and their own large fisheries using mother ships and factory ships. This strategy is particularly obvious in the case of the Japanese regulatory system under which fishing grounds are allocated among different sectors of the industry in order to minimize conflicts between them. Thus, mother ships accompanied by catcher boats, as well as large factory ships, are not allowed to operate in waters close to the home islands and are pushed further and further to distant-water grounds, most of which (except for the tuna longlining grounds) are close to the coasts of other nations.

For these reasons and others not necessarily founded on rational calculation, it is understandable that the growth in distant-water fishery exploitation has had appreciable effects upon international fishery regulation. The real or imaginary impact of these fisheries is a major factor in motivating unilateral jurisdictional claims by coastal states. In the North Pacific, unilateral actions have already been taken by the United States, Canada, and (more than twenty years ago) by South Korea. Even the Soviet Union has taken such an action to protect its fisheries against Japanese high-seas activities. In the world as a whole, extensions of fishery jurisdiction are the rule, not the exception, and the expansion of Japanese and USSR distant-water efforts are partially, but very importantly, responsible for this worldwide trend.

We consider the distant-water/coastal state conflict as a contemporary problem in the North Pacific because, despite the actions already taken by coastal nations, further measures to seek additional protection can be anticipated. There is continuing agitation within the United States and Canada for broader jurisdiction over living resources or for other methods of acquiring greater rights for coastal fisheries. Sooner or later, perhaps in connection with the general LOS negotiations, these two nations are likely to take further actions toward this objective. At the moment, at any rate, the problem is still unresolved despite current agreements.

PROMPTNESS OF ACTION

It is no longer uncommon for a large fishery based on a new resource to develop quickly, sometimes within two or three years. Various strategies must be incorporated into the management of fisheries in the North Pacific to permit prompt action in these situations. These include a system for monitoring developments and efficient and compatible national statistical systems, coupled with an obligation to report data promptly to the international monitoring authority.

Another shortcoming, which afflicts the North Pacific and many other regions in the world, is that management decisions or recommendations often have to be based on the kind of evidence that requires lengthy periods of time to produce, if it is possible to do so at all. It is common to many of the existing fishery commissions that parties demand, as is permitted under their charters, that regulatory actions be supported by indisputable evidence. This philosophy, which essentially rests on the unwillingness of states to allow effective third-party decision making, must be changed if international action is to have a desirable impact on fishery management. One suggestion now commonly advanced is that decisions or recommendations should be based on the best available evidence. Another, less drastic but still a substantial departure from the need for proof beyond reasonable doubt, would be to permit action based on the preponderance of the evidence.

The length of time that elapses between the adoption of a recommendation and its enforcement is another factor in effective management. Although this period has become prolonged in many international bodies around the world, the North Pacific bodies work rather well in this respect, perhaps because they were created ad hoc to reflect the political and other realities of specific situations. Once agreed upon, such regulations as catch quotas, closed seasons, closed areas, and limitations on fishing effort are put into force promptly in most cases. In constructing a future regime for the North Pacific, it is important to retain this feature in an overall fishery organization.

COST OF ADMINISTRATION

Another problem of existing fishery arrangements around the globe—one that is particularly obtrusive in the North Pacific—is the large amount of time and effort expended, and mostly wasted, on routine meetings and research discussions. The number and frequency of meetings has proliferated so much, and both administrators and research workers concerned with international fishery problems are so fully occupied with preparatory work, that practically no time is left to think about long-term management problems or to engage in more productive work such as the development of new resources. This waste of valuable time is even more regrettable because many of the discussions at regular meetings are static in nature: nothing much can happen unless some political agreement is reached at higher levels.

V Negotiations on the Law of the Sea And Fishery Management in the North Pacific

IN CHAPTER III we identified outstanding problems of fishery management in the North Pacific. In this chapter we will assess the prospects for remedying these problems through the general outcome of the third conference on the law of the sea.

There can be no reasonable doubt that issues concerning fishery allocation and conservation, embraced in the general term "fishery management," are central in the current negotiations on the law of the sea (LOS). Many developing states consider that questions about living marine resources are the most important of the numerous LOS issues, and some developed states, especially the Soviet Union and Japan, rate these matters very high among national priorities. Even the United States, whose domestic fish harvesting industry has not generally progressed as well as the rest of the economy, has indicated that fisheries issues have considerable priority, though they are, perhaps, subordinate to other interests.

It is expected, accordingly, that there will be numerous proposals and much debate on fishery matters at the third LOS conference.[1] If the conference is postponed, fishery issues will probably be more rather than less prominent. Even at this early stage of the proceedings (fall of 1972) proposals have been tabled by several states and others are to be anticipated. Debates thus far, although not extensive, have clearly indicated that there are potentially wide differ-

[1]After this was written, the General Assembly of the United Nations voted to request "the Secretary-General to convene the first session of the third UN Conference on the Law of the Sea. . .in November/December 1973, for the purpose of dealing with organizational matters," and decided "to convene the second session of the Conference, for the purpose of dealing with substantive matters, at Santiago, Chile, in April/May 1974, for a period of eight weeks. . . ."

ences of opinion on fisheries issues and these differences may grow sharper as discussions become more detailed.

Alternative possibilities for LOS negotiations include widening the territorial sea to a substantial extent, maintaining a moderately wide territorial sea coupled with a much larger fishing zone, and a variety of forms of extended coastal authority delimited in a number of different ways. We will attempt to analyse the probabilities of acceptance and the potentialities of these various alternatives for resolving the outstanding substantive management problems confronting North Pacific fisheries.

PROCEDURAL FACTORS

Certain procedural factors will influence the effectiveness of the third LOS conference in its attempts to resolve the difficulties in this region or elsewhere. An initial consideration concerns the conditions for reaching agreement at the conference. Under the usual UN procedure, any substantive provision must command a two-thirds majority of those present and voting for incorporation in any treaty formulated by the conference. Mere majority approval is insufficient. Accordingly, effective opposition by one-third plus one of those present and voting during a meeting will prevent adoption of a proposal. Although a simple majority of states might thus wish to govern their relationships by a particular regulation, it will not be possible to achieve this through the next LOS conference if it employs the usual voting procedure for UN conferences.

Even if a provision is adopted by a two-thirds majority, it does not follow that its effective implementation is assured. Any treaty so approved by the conference must first be ratified by an agreed-upon number of states before it comes into force between the ratifying states. This quite customary step is not likely to be much of an obstacle since the number of ratifications required is not usually large. In this instance, the number of ratifiers needed to bring a treaty into force is likely to be quite small in relation to those participating in the conference itself.

If a treaty is to be effective, a far more important condition is the composition of the states ratifying or adhering to it. In connection with fisheries, the significant point is whether major participants in particular regions and fisheries become parties to agreements pertaining to or affecting this activity. The 1958 Convention on Fishing and Conservation of the Living Resources of the High Seas was largely ineffective because of insufficient acceptance by important fishing states. The "special interest" of the coastal state and the superior rights flowing from this interest, as recognized in the convention, were left unimplemented when distant-water states refused to adhere to the agreement. Thus, directly pertinent experience demonstrates that two-thirds approval at a conference does not signify an effective treaty.

Accordingly, if the third LOS conference merely succeeds in getting two-thirds approval of agreements concerning fisheries, it will not necessarily have accomplished much. This is seen most dramatically, perhaps, in light of the relatively few states which account for the bulk of the world's fish catch. The top six states took about 40 million metric tons of the total 69.3 million metric tons caught in 1970. The overwhelming preponderance of the remaining 29.3 million metric tons was caught by twenty-five states. If a fisheries arrangement is not ratified by a large portion of these thirty-one states, it will not matter very much whether the agreement is enthusiastically accepted by a substantial body of the remaining states. Nonratifiers would not feel bound by provisions of the agreement, however many other states ratify it, and would consider action pursuant to the treaty as unilateral.

Consequently, it appears that the probabilities for an effective agreement concerning fisheries may not be especially high for any region in which major distant-water fishing states are involved, if, as seems likely, the conference adopts treaty provisions strongly protective of coastal fishing interests. This bears especially on the situation in the North Pacific because the two foremost distant-water fishing states are Japan and the Soviet Union, and both focus a very large portion of their total activities in this area. Japan, for example, derives an estimated 70 percent of its distant-water catch from two parts of the North Pacific—the Kamchatka-Kurile area and the eastern Bering Sea. In view of the considerable negative impact that the third LOS conference might have upon Japan and the Soviet Union, it would not be surprising if these states declined to accept agreements on fisheries.

However, the attitude of states toward agreements ensuing from the conference will depend on many factors, including the relationship between various agreements and the form in which a particular agreement is embodied. Be this as it may, the substance of the agreements could be, we assume, of decisive importance; hence we will examine some alternatives now being proposed. We will then consider the relationship between these alternatives and the major problems previously identified as outstanding in the North Pacific region and attempt to estimate the degree to which the alternatives could resolve or lead to the resolution of such problems.

Coastal nations feel most strongly that a solution to fisheries problems lies in control over fisheries in the ocean off their coasts. The degree of this interest is reflected in the wave of extensions of national jurisdiction, in a variety of forms, over the decade of the 1960s and, more recently, in the type of proposals being discussed most prominently in the preparatory meetings for the third LOS conference. The record of the former development is amply clear but it can be emphasised that the vast majority of coastal nations have made very modest extensions of their authority to dispose of fish adjacent to their coast. As of late 1972, only 20 of approximately 147 political entities in the world claim control over fisheries beyond 12 miles. However, it is clear from international discussions to date that many more states wish to acquire greater control through one method or another of extension. We can deduce that there is widely felt apprehension (genuine and spurious) over the pressure of distant-water fishing on adjacent resources and that many states will seek a means for dealing with the situation through agreement on extended coastal controls over fishing at the third LOS conference.

ALTERNATIVE PROPOSALS

Ideas and alternatives under discussion for achieving the protection that coastal nations desire or demand include: wide margins for the territorial sea, a modest limit for the territorial sea coupled with recognition of coastal authority over fisheries in a zone beyond the territorial sea, recognition of coastal control or rights beyond the territorial seas over certain specific fisheries or stocks, delegation to coastal states of jurisdiction over fisheries coupled with establishment of internationally agreed-upon criteria for allocation or conservation, and some combination of these. Another means for expanding national jurisdiction could be to establish a new definition of living resources subject to the existing Continental Shelf Convention or to a new seabed treaty.

An extension of national territory through the establishment of a very wide limit for the territorial sea would, in accordance with prevailing law, permit the coastal state to exercise exclusive and absolute control over any living (or nonliving) resource found therein. This alternative presents formidable difficulties, however, and it does not at present seem a likely outcome. A substantial number of nations appear to prefer a narrow territorial sea in order to minimize potential hazards to important nonextractive uses of the sea, particularly the movement of vessels and aircraft. It is unlikely that coastal states would in fact take unilateral action to severely restrict a right of passage for nonmilitary purposes within their territorial sea (including straits) because practically all states benefit directly from this right and also because such hostile action could trigger retaliatory countermeasures of various kinds. Although a proposal for a territorial sea broader than 12 miles is highly improbable, with such a big stake there would probably be enough nations concerned to block it. In sum, the chances seem slim for an effective global treaty specifying a territorial sea significantly broader than 12 miles to emerge from a LOS conference.

Any effective global agreement on fishery matters that does result from the conference will probably be based on the concept that authority over fisheries can be separated from the total package of national jurisdiction comprising sovereignty. Such a procedure was attempted in 1958 and 1960 and barely failed. Even if this approach fails again, it does seem likely that the conference will provide evidence of a general recognition, by a large majority of nations, of a need to provide for special rights of coastal states either in terms of exclusive fishery jurisdiction or other forms of preferential allocation of resources.

Provisions for recognition of coastal authority over fisheries might take a number of forms and vary in terms both of degree of authority to be exercised by the coastal state and of the delimitation of the area to which authority applies. With respect to the degree of authority, the sharpest relevant distinction is between coastal control over access to resources and coastal authority to provide for management without allocation. The best examples of the former, perhaps, are the familiar exclusive fishing zones that have been proclaimed by many states at one time or another. In these zones the coastal state is regarded as having the same degree of control over living resources as it has in its territorial sea; i.e., a complete power of disposition. A coastal control limited solely to management is often understood to mean (and is so understood in this discussion) that the coastal state may control rates of use or set limits on use but is not authorized to determine who may use; i.e., may not allocate resources. The 1958 Convention on Fishing and Conservation of the Living Resources of the High Seas exemplifies this approach.

Coastal states' authority over disposition need not be complete or absolute and some current thinking does envisage a lighter degree of control. Thus, another way of protecting coastal interests would be to confer a preferential right to certain or all resources of an area beyond the territorial sea. Under such a system, a certain portion of a particular stock or stocks would be allocated to the coastal state, but foreign fishermen would continue to have some access to stocks not used, or only partially used, by the coastal state. It is obvious that the criteria for coastal preferential rights are of the highest importance, as is the determination of who establishes and applies such criteria and settles disputes. A wide variety of arrangements could be made to accommodate coastal and distant-water interests. In extreme forms, provision for preferential coastal rights might not differ from a straight exclusive fishing zone, although some proposals apparently envisage less than this. It is possible, for example, to conceive of a general agreement that would lay down the general criteria determining the existence of a preference for the coastal state, and also provide for a third-party method of applying such criteria or of reviewing decisions by the coastal state on their application.

The means for delimiting the area of application for such rights as are conferred on coastal (and other) states also vary a good deal. A simple, but highly controversial, method for handling this is merely to define a zone in terms of distance from shore. This is perhaps the most familiar approach since so many states have already established "zones" that are bounded by artificial lines at specific distances from the coast. However, some have perceived a danger that a zone so delimited may become transformed over time into a territorial sea over which the coastal state exercises far greater authority.

Other suggested alternatives for delimitation include depth. The close relationship between living resources and the continental shelf is apparently responsible for these suggestions since the shelf is often defined in terms of depth.

Still another method sometimes preferred would define coastal control in terms of species or stocks categorized by their association with specific areas or, in part, by their life history. (This combination is already an accepted method for certain species with a particular association with the ocean bottom.) In this method the coastal state would have designated authority over most stocks occurring in waters above the shelf or slope as well as over those that spawn in fresh water and then live part of their lives at sea. A different management system would be provided for oceanic migratory stocks. U.S. Article III, tabled with the UN Seabed Committee in August 1971, contains one form of this method of delimitation.

This method for delimiting control over certain resources by defining them as continental shelf resources might also be employed to include stocks with less physical connection with the shelf than more or less continuous direct physical contact. The effects of this technique could be quite substantial. For example, Article 2 (4) of the Shelf Convention could be amended to broaden the definition to include such forms as shrimps and flounder as subject to sovereign rights for the purpose of their exploitation. Such a provision would provide ground for prohibiting foreign trawling activities at least within the 200-meter depth contour and very probably beyond.

DIFFICULTIES OF ACCEPTANCE

All these alternatives confront formidable obstacles to acceptance in the forthcoming negotiations. There are, first, different ideas about the best means for protecting coastal fishery interests, although the general principle of such protection seems widely accepted. There are also opposing ideas about the way to define the area within which coastal rights (whatever they may be) are paramount. Matters are further complicated because these two issues are sometimes evaluated separately. Thus, even states in accord on the scope of the substantive rights of coastal states over fisheries may disagree on the method for delimiting the area to which these rights apply. Furthermore, even if they agree on the delimitation method, they may disagree on the exact numbers to be used in its application.

It is possible that a great many states now, or will, share the view that each coastal state should have an exclusive fishing zone within which it has the exclusive right to fish and to regulate all fishing that takes place, no matter by whom. Despite this, areas of substantial disagreement could still exist or arise. States placing the same general and high priority on their coastal fishing interests could differ in the weight they give their distant-water interests. They could also differ on the criteria for protection of historic fishing rights within whatever formula is adopted for safeguarding coastal interests. For example, this difference in appreciation of distant-water interests might affect the respective views on how wide an exclusive fishing zone might be. In one case, a 50-mile zone (including or beyond a territorial sea of 12 miles) may offer ample protection for coastal fisheries but by reciprocal adoption pose a threat to the coastal state's distant-water fleet. In another instance, a coastal state may believe that it requires at least a 100-mile zone and that its distant-water fishermen will be able to live with that limit too.

These disagreements about the size of an exclusive coastal fishing zone are sharper and more easily detected than differences over other possible features of a regulatory system. Arrangements for allocating preferential rights within a fixed zone off a coastal state are potentially far more complex and specific formulae are shaped differently from one situation to another. If preferential rights pertain to specific resources rather than to an area, then the situation is still more complex and a number of problems arise in their implementation.

The treatment of anadromous fisheries, as well as of various marine mammals returning to land for breeding, will probably be considered a special problem for general negotiations. As previously noted, different forms of allocation are currently employed by nations in the North Pacific for handling these resources and still other arrangements apply elsewhere. It is clear that international arrangements for these particular animals cannot simply be subsumed under the provisions dealing with coastal and high-seas pelagic species. Although this problem is not yet prominent in the preliminary LOS negotiations, it will undoubtedly play an important part as time goes by.

The important question for purposes of the immediate discussion is how these various alternatives meet the outstanding problems in the North Pacific, both in a substantive sense and in the sense of the political realism of potential effectiveness.

It seems apparent that on either or both counts there is no great reason for optimism that the third LOS conference will resolve the fishery management problems of the North Pacific. Let us look first at the political chances of effective agreement.

At the present stage in negotiations there seems to be a strong base of support for agreement on a territorial sea of 12 miles, coupled with a zone beyond, probably up to 200 miles, within which a coastal state would have exclusive fishing rights. The third LOS conference may succeed in achieving a two-thirds agreement on these or essentially similar provisions. Some North Pacific states are almost certain to refuse to accept such a treaty, including Japan, South Korea, and perhaps the Soviet Union. (Unless the transit issue and anadromous fish problem are also resolved, even the United States might not accept this agreement.) If these three states do refuse to ratify, practically all existing problems in the North Pacific will continue. Even if Japan alone refuses to accept, major questions between North Pacific states will still have to be resolved in some other fashion; for instance, by unilateral action, further negotiations, or even litigation.

The political acceptability of preferential rights in the North Pacific region might be somewhat higher than an extensive out-and-out exclusive fishing zone. It would seem that, extraneous factors aside, the key determining factor here is the degree of impact on distant-water fishing. If the coastal rights are too extreme, if foreign fishing is excluded too easily, or if the costs of gaining access to coastal waters are excessive, then there is no great reason to expect adherence by Japan, South Korea, or perhaps the Soviet Union. Again, in these circumstances the present problems continue to exist and some further negotiations will be needed.

CONTINUING NEED FOR NEGOTIATION

Even if the next LOS conference agreed on treaty principles recognizing certain preferential rights and these were accepted by all the North Pacific states, there would probably still be a need for further negotiations. Such an accord would perhaps be in general terms or, at least, in terms that would require some judgment or discretion in application to concrete circumstances. In either of these situations, specific states will probably negotiate further on (1) establishing the basis for preferences; (2) applying the general criteria for preferential rights to specific circumstances; (3) determining the amount, kind, place, and method of foreign fishing; and (4) perhaps establishing a mechanism for routinizing the handling of these questions and others as they recur from year to year.

In sum, if these negotiations do occur they must confront many of the same problems, especially of allocation, that are already outstanding in the North Pacific arrangements.

A possible, but not likely, outcome of the third LOS conference could be a convention providing for general principles of conservation and allocation applicable in appropriate circumstances to any fishery. Even though such a treaty might be acceptable to most of the major fishing nations in the region, they would still have to make specific arrangements in light of specific but varying conditions. Such an agreement could not, therefore, move us very far beyond the immediate situation since existing agreements already take into account in varying degrees the need to protect the interests of coastal fisheries.

Still, such an agreement would have the advantage of lending greater authority to the principles enunciated. Furthermore, some general principles might assist in the development of better management and information systems in the North Pacific and elsewhere. In addition to the need for protection of coastal fisheries,such principles or guidelines might include, inter alia, the necessity to deal squarely with the question of allocation wherever applicable; the obligation of each nation fishing in a region to collect and to submit catch and effort data, as well as certain types of biological information in a uniform format (an obligation applicable on a worldwide basis for certain species, such as whales and tuna); the establishment of panels composed of scientists from nations not directly involved in order to obtain independent judgments on scientific issues; and the need to develop a system of international enforcement, at least to the extent of providing for mutual inspection.

If an agreement of this type cannot be reached in current LOS negotiations, it is still conceivable that these principles will become generally acceptable and applicable in various parts of the world through evolution of management practices. Thus, their acceptance in the North Pacific region and the creation of better management institutions might assist regulation of fisheries not only in this region but also in the long run, elsewhere in the world.

NEED FOR IMPROVED INSTITUTIONS

No matter what is agreed at the third LOS conference on substantive principles of fisheries management or on fishing zones, there is almost certain to be a continuing need for improved institutional mechanisms in the North Pacific. Up to now, practically all international LOS discussions regarding fisheries have concerned exclusive or preferential rights and have ignored better institutional structures, except for very general reference to international mechanisms as in the original U.S. Article III. None of the principles or zones being discussed would remove the need for such structures. Any agreements that do occur are almost certain to leave issues outstanding that require additional, perhaps even more intense, negotiations or bargaining within this region and probably in others. If this happens, it would be convenient, and perhaps even necessary, to have an institution to facilitate the economic bargaining and communication required for handling the complex problems of a great ocean region.

One potential, and the most probable, outcome of the third LOS conference has only been mentioned in passing; that no agreement at all may be reached and that fisheries issues might be left unresolved. Such a failure to agree would have reasonably obvious consequences: an increasing number of states would take unilateral action of one form or another to extend their authority over fisheries.

Let us first consider the reaction of states whose fisheries could be affected by extended fishing claims. From a review of the numerous unilateral extensions in recent years, it can be seen that the responses of distant-water fishing nations vary greatly. In many instances the distant-water states have voluntarily refrained from fishing in the zones unilaterally established by coastal states. In other instances the foreign fleets have continued fishing at the risk of seizure of their fishing craft, while simultaneously filing official protests. Where the new jurisdictional claims have not been extensive, negotiations have produced various types of accommodations. Sometimes arrangements have been made to phase out the distant-water fleets within a specified period of time. Occasionally short-term agreements have allowed continuation of the distant-water effort subject to renewal or revision of the agreements.

Other types of arrangements for permitting foreign fishing include payment for fishing in the claimed zone, financial and technical assistance, or use of local facilities. Some agreements, such as the North Pacific bilaterals between the United States and other states noted earlier, are on a give-and-take basis. Some foreign fishing and other related activities are allowed to continue in certain areas within the claimed zones but are prohibited in some areas outside in order to reduce adverse effects on coastal fisheries. Sometimes neighboring countries claiming extended fishery jurisdiction make reciprocal arrangements to accommodate each other's fishing activities within the respective zones. As already noted, these agreements are usually for short periods and require frequent renewal or revision.

The most important feature of the whole pattern of reaction to unilateral action is that, except for isolated instances, such claims have not been challenged by force. The future does not seem likely to be greatly different in this respect unless the unilateral claims seek to affect more than just fishery operations. Distant-water and coastal fishing states will continue to resolve their differences through negotiations.

Indeed, it is realistic to expect that future changes in the international fishery regime, which will probably continue to exhibit the currently predominant trend toward extension of national jurisdiction, will result in more international negotiation rather than less. We have already noted that adoption of any one of the various alternatives now considered realistic would still leave a number of important matters for subsequent negotiation. We have not discussed the whole question of resolving the boundary problems that will emerge as states continue to extend national controls over fisheries further out from land. It will also be necessary for neighboring states in many regions to negotiate arrangements for accommodating each other's fishing activities. Without a deliberate effort in this direction, the development of the fisheries of coastal states will be seriously hampered and proper management of fish stocks crossing several national boundaries will be impossible. Consequently, in many areas a complex network of bilateral, subregional, and regional agreements will emerge for dealing with the problems caused by expanding national fishery claims.

This system is not an efficient or effective means for dealing with management problems. It certainly leaves much to be desired in the North Pacific. The existing regime consists of a variety of uncoordinated conventions and agreements concluded to resolve specific issues, mostly of a political nature. There is a definite need for an overall agreement to coordinate, and perhaps eventually to replace, the existing arrangements, or at least some of them. As just indicated, the present and prospective law of the sea negotiations will probably intensify this need rather than obviate it.

A new institutional arrangement in the North Pacific should have several features, including that of being open to new entrants. An essential requirement is a built-in monitoring system with each participant accepting the obligation of prompt submission of statistical and biological information according to established formats. On the other hand, the new regime must maintain, and where necessary establish, some features of the existing arrangements, particularly in directly confronting and resolving problems of allocation. The increasing need to accommodate the interests of coastal and distant-water states deserves special attention. There is also a need for more realism about the nature of the scientific evidence used as a basis for decisions to implement management measures. The promptness in implementing agreed-upon measures which, in part, characterizes the present system should be carried over. Particular effort should be made to reduce the amount of time and effort uselessly devoted to routine meetings, thus freeing researchers and administrators for more creative and productive work.

We will now turn to the way in which the transition might be made from the existing agencies and arrangements to a future agency, and will outline the salient features of this agency.

V Alternative Arrangements for a North Pacific Fisheries Management Agency

HOWEVER INSUFFICIENT THE PRESENT CONGLOMERATION of fisheries agreements and practices is for effective management in the North Pacific, their existence is a major political fact that must be recognized in attempting to improve the situation. Such recognition means that, for all practical purposes, these arrangements cannot be terminated immediately; rather the basic philosophical approach must be one of gradual transformation. This is an important factor in the structure of a new institution for the North Pacific.

First we will examine more important existing arrangements in the North Pacific and discuss whether and how they may be modified to fit an overall institutional system. It will be seen that, except for salmon (and crabs to a lesser extent), the major living resources in the North Pacific do not present any insurmountable problems in assimilation within a new framework. However, this relatively happy situation will prevail only if action to initiate the transformation occurs fairly soon, before major disputes develop from exploitation of resources at present not subject to international regulation.

EXISTING ARRANGEMENTS AND WAYS TO TRANSFORM THEM

We have already demonstrated the complexity of the existing pattern of agencies and regulations. Their transformation into a regime is likely to be a complex process. Each individual arrangement will now be examined, beginning with those that seem relatively easy to handle.

Whaling Convention

Problems of whaling exploitation are the subject of the International Convention for the Regulation of Whaling, which applies throughout the world ocean, including the North Pacific. Since whaling problems should be considered and resolved in their total global context, it is sensible to exclude whales from the resources to be managed by a new North Pacific agency.

Tropical Tuna Convention

The existing arrangement (The Inter-American Tropical Tuna Convention) continues to be an acceptable mechanism for handling tropical tuna problems and there is no reason to add these problems to those confronting North Pacific nations and arising in the region. The tuna commission might in the future broaden its base by expanding its coverage in terms of both area and species. Eventually there should be, and probably will be, a tuna management system applicable on a worldwide basis, including tuna caught in the North Pacific.

Fur Seal Convention

The present system for managing fur seals (The Convention on The Conservation of North Pacific Fur Seals) might well be maintained intact and separate from the new agency if it were not for the possibility, admittedly small, that a new nation from the region might begin pelagic sealing. If this happened, it would be desirable to have a new treaty open to all nations desiring to participate, setting forth principles for handling the matter of new entry.

Applicable principles would continue the prohibition against pelagic sealing and maintain basically the present system for distribution of shares. The most important new principle would be that a member state other than a party to the existing fur seal convention could participate in the profit-sharing system, if it made a capital investment that would yield an interest equivalent to the value of pelts or proceeds to be distributed. The amount so invested would constitute an international fund, the interest from which would be used or distributed according to an agreed-upon formula. It would probably be easier to conclude such an arrangement since all existing sealing enterprises in the North Pacific are governmental or public in nature, and there are no private groups with fixed expectations of gain. We doubt that a new nation would actually wish to participate in the sharing system because of the requirement of cumbersome international negotiations and new administrative procedures established for this purpose. But the possibility is still sufficient to warrant the new arrangement and its inclusion in the new agency.

If a nonparty to the new agency seeks to undertake pelagic sealing, thereby threatening the conservation and distribution system among the parties, it could of course be invited to participate in the treaty. No doubt such a contingency would involve political negotiations and settlement among all those concerned. The continued successful operation of the fur seal treaty for about sixty years strongly suggests that there is no great prospect that a new state will enter to disrupt the arrangement.

The whole question could, of course, be handled by agreement among the parties to the existing convention to amend it to include the provisions outlined above. The advantage in incorporating the new fur seal arrangement in a new agency is that negotiations would be permitted among the agency's members, probably in the form of a special panel for fur seals, and all North Pacific arrangements could be consolidated in a single comprehensive structure.

Fraser River Salmon Convention

Unless the existing system of protecting North American salmon stocks[1] breaks down so that high-seas salmon fishing becomes a real threat to the conservation and allocation system now applicable to the Fraser River stocks, little would be gained by merging the salmon fishery with a new North Pacific agency. It has been demonstrated that very few, if any, Fraser River salmon are caught by Japanese fishermen west of the present abstention line. We do not anticipate any prospect that the protection system will fail; hence it is better to maintain the present commission as it is, especially since it deals largely with management problems wholly within the limits of national jurisdiction.

East China Sea Fisheries

The major participants in fishing in this subregion are Japan, North and South Korea, and China. Until political relations among these parties improve, there is little that can be done to upgrade fishery management for the region's fisheries. Despite these political complications, the new agency should have a panel for these fisheries, with participation by any members directly concerned with exploitation of the demersal and pelagic resources in the area. Pending establishment of a regulatory system for this subregion

[1]The Convention for the Protection, Preservation and Extension of the Sockeye Salmon Fisheries of the Fraser River System (1937).

embracing all parties concerned, the panel should take up such matters as its members consider important.

We now turn to the more difficult issues, including arrangements for high-seas salmon fishing, high-seas crab fishing, high-seas trawl fishing, and the problems arising from extension of coastal jurisdiction.

Salmon

Although the regulatory systems in the eastern and western Pacific are entirely different, the four major salmon fishing states in the North Pacific concur in giving high priority to salmon problems among their fishery interests in this area. The importance attached to salmon fishing by these four possible parties to a new treaty is sufficiently large that any drastic change in the status quo is likely to provoke strong objection by one or more of them.

The most drastic course of action one can imagine is the complete prohibition of high-seas salmon fishing, the claimed justification being the special rights of states possessing spawning streams to salmon management and harvesting. This probably would not be acceptable to Japan under any circumstances. It is possible that the other three states might then wish to take coordinated unilateral actions against Japan, but such actions would create strong negative incentives for Japan to participate in the new convention. Chances seem only slightly better for Japanese acceptance of such a principle when coupled with a very long phase-out period.

Another possible way to deal with this difficult problem is a gradual change in the pattern of high-seas salmon fishing on the grounds of the need to manage resources better, but to make it in a way that still takes into account the historical interest of Japan. For example, instead of establishing extensive closed areas in waters near the coast, high-seas fishing could be allowed in limited areas fairly close to the coast so that the origin of the fish taken is better known to the coastal states managing the resources. This would also reduce, to a great extent, the catch of immature salmon. A quota system could be employed, with catch limits set on the basis of conservation needs and historical interest. Such an arrangement would also encounter a variety of difficulties from its impact on the interests of various states. For Japan, the new arrangement would require a major realignment of the licensing system for all salmon fisheries. Also, since the total effort required to take the same amount of fish would be much smaller under the new arrangement, many vessel owners and fishermen would have to be excluded from salmon fishing. If, however, the new convention could accommodate a gradual transformation over an extended period of time, the basic idea might not be rejected by Japan.

It is difficult to speculate on the response of the Soviet Union to such a proposal. The new arrangement would be a substantial improvement over the present situation from the point of view of management of USSR salmon stocks; but the Soviet Union, too, might have internal political problems.

The United States might find the proposal unacceptable if Japan insisted on being entitled to catch a certain amount of salmon of North American origin, particularly Bristol Bay sockeye salmon, on the basis of an historical interest. While Japanese fishermen have been taking substantial amounts of sockeye salmon of Bristol Bay origin, the United States would argue that Japan has been able to do so only because of a lack of agreement on the location of the permanent salmon abstention line. Basically, Canada would have no reasons to object to the proposed new arrangement except for possible repercussions from fishermen in British Columbia.

A number of technical problems are presented by this arrangement. These, however, would not be as difficult to solve as the political problems, for fundamentally the new system would be an improvement over the existing one. The question of new entry, in this case, would have to be handled on the basis of historical interest. In other words, members joining the new convention who had not fished salmon on the high seas for a certain number of years in the past would have no right to initiate such fishing. This is in effect not different from the status quo.

Some might wish to carry innovation a step further by introducing a system of revenue distribution (in terms of either fish or proceeds). The system would differ somewhat from that for fur seals, since Japan would be the only recipient of fish or proceeds. One might go even further by proposing that Japan buy salmon from the Soviet Union, and possibly the United States, at reduced costs instead of receiving them in return for giving up offshore fishing. Either proposal would be difficult for Japan to accept because it would mean almost the complete elimination of the Japanese salmon fishing industry except for its processing and marketing sectors.

A third choice would be to maintain the status quo by simply transferring the relevant provisions of the North Pacific treaty and the Japan-Soviet Union treaty to those under the new convention. From a purely political point of view, this is an attractive way to approach the problem. But from the point of view of conservation, the new convention would inherit all the shortcomings of the present arrangements. In addition, the new system would still have to face the question of new entry unless the convention included a provision to prohibit high-seas salmon fishing by a nation that had no historical interest.

Perhaps the most practical course of action is to start with the third alternative (status quo) and gradually move into the second one (fishing in limited areas relatively close to the shore) and, if feasible, to a system of revenue distribution. This is a slow and rather clumsy process, but it would be toward better management of salmon resources in the region. There is, in any event, no way of handling this matter without causing substantial difficulties for one nation or another. But the fact remains that, unless the salmon arrangements are acceptable to the four nations, there probably will be no overall convention in the North Pacific.

Crabs

The high-seas crab fisheries are much less important to the various fishing states than are the high-seas salmon fisheries. Despite contradictory views on the legal status of this resource, the states concerned have managed to reach a series of agreements containing tighter and tighter controls for the purpose of conservation and allocation. The species coverage of these agreements has also been expanded from king crab to those of most of the other important species of crabs in the high seas of the northern North Pacific. In addition, both the Soviet Union and Japan have refrained from developing crab fisheries in waters south of the Aleutian Islands and the Alaska Peninsula. The present situation thus already reflects, by agreement and otherwise, de facto recognition of some special rights of coastal states, as well as some consideration for the historic interests of noncoastal states. A new convention might be more explicit about this. Since some of the most important crab stocks have been depleted, presumably by overfishing, the conservation and restoration of these stocks should be given special emphasis by the new agency.

Japanese common name	English common name	Scientific name
Tarabagani	King crab	*Paralithodes camtschatica*
Aburagani		*P. platypus*
Hanasakigani		*P. brevipes*
Zuwaigani	Tanner crab	*Chionoecetes* spp.
Ibaragani		*Lithodes aequispina*
	Dungeness crab	*Cancer magister*
Kegani		*Erimacrus isenbeckii*

Herring

In many areas of the western Pacific, herring resources, which were once very large, are in very poor condition and show no sign of recovery. The important Asian herring stocks are now found in the northern Okhotsk Sea, along the west coast of Kamchatka, and along the Siberian coast of the Bering Sea. In addition, the Soviet Union and Japan have developed substantial fisheries in the central and eastern Bering Sea, the combined catch having reached 140,000 metric tons in 1970. Herring found in the Gulf of Alaska and areas further to the south generally stay in coastal waters (more or less within the present limits of national jurisdiction) throughout their life cycle, with some exceptions, and therefore are not likely to become sources of international conflicts. Japan once made a serious attempt to develop a high-seas herring fishery in waters around Kodiak Island but failed completely.

Only two international agreements affect herring in the North Pacific: a Japan-Soviet Union agreement as part of the Japan-Soviet Union Convention Concerning the High Seas Fisheries of the Northwest Pacific Ocean, and the abstention principle as applied to herring populations in southern British Columbia as part of the North Pacific fishery treaty. The herring problems of the North Pacific, therefore, could be treated within the general framework of the new convention on the basis of conservation, and, if necessary, national quotas, making allowance for new entrants and for preferential allocation to coastal fisheries.

Trawl Fisheries

Among the major fisheries currently not regulated internationally, potentially the most serious international problems are likely to come from the trawl fisheries in northern areas, including the Bering Sea, the Gulf of Alaska and further southward, and waters off Kamchatka, the northern Kuriles and Hokkaido-Sakhalin. The exploitation of pollack stocks in the Kamchatka-north Kurile area and the Bering Sea is likely to become a very controversial issue because the catch appears to be close to the maximum sustainable yield; and the Japanese, Soviets, and Koreans are all expanding their activities. Few of these trawling activities are at present subject to international regulation, except for those subject to bilateral agreements resulting from extension of national jurisdiction and those affecting the halibut stocks which are subject to regulation by the INPFC and, for the United States and Canada, by the

Halibut Commission as well. Future problems concerning these fisheries, therefore, are definitely candidates for treatment within the general framework of the proposed new convention.

Extension of Coastal Jurisdiction

As previously discussed, continued expansion (in most cases unilateral) of national jurisdiction over fisheries and perhaps over other activities is likely. However, there are good reasons to doubt whether such extensions would solve most of the problems of fisheries management in the North Pacific, nor would extension of national limits over fisheries result in less international negotiation over fishery matters. However, some North Pacific nations may consider extended coastal jurisdiction to be the best solution to fishery problems; and, if so, consideration should be given to what a new management agency might do to approach problems arising therefrom.

An initial fundamental point is that this general trend toward extended national jurisdiction will result in underutilization of some large fishery resources unless coastal states are willing to accommodate foreign fishing for resources that are not utilized, or exploited only to a limited extent, by their own fishermen. There are, of course, many ways for coastal fishing nations to accommodate foreign fishing without serious adverse effects on their own fisheries and even to derive substantial benefit from such fishing, including the issuance of licenses, users fees, and other methods of producing income. But the general public of the coastal state may categorically oppose any foreign access to national jurisdiction, whether or not the coastal state is in a position properly to utilize living resources in waters under its control.

SALIENT FEATURES OF THE NEW AGENCY

Before examining specific components of a potential new set or sets of machinery, which we label for present purposes as the North Pacific fishery agency, some thought may be usefully directed to the general framework of such an organization. Of the variety of possible forms it might take, some can be eliminated at the outset.

First, very few, if any, fishery people now consider a worldwide regulatory agency as a realistic or even a reasonable proposition except for certain resources such as whales or some species of tuna that require regulation on an interregional or worldwide basis because of their biological characteristics. Most international fishery problems in the North Pacific that are likely to arise in the foreseeable future are amenable to treatment by, at most, seven or eight sovereign states. There is no discernible advantage in involving one hundred or more sovereign states in one way or another when trying to settle these problems.

Another form of organization not likely to be supported by the major fishing nations of the North Pacific is a regional fishery committee or council for this region of the Food and Agriculture Organization of the United Nations (FAO). None of the existing FAO councils or committees is functioning as a regulatory body. The Soviet Union is not even a member of FAO; and some of the other nations, particularly the United States, have not favored FAO involvement in international fisheries management through its committees. Although for these reasons an FAO involvement in management seems unlikely, there may still be a role for FAO to play in the initiation of a new agency, a point mentioned further below.

Establishment of the Organization

Timing. At present, a major obstacle to any initiative affecting any matter connected with the law of the sea is the pendency of the third LOS conference on the subject and an overwhelming concern about its outcome. Apparently every move by a major maritime power is assessed, internally as well as externally, in terms of its possible effects on these negotiations and the result is to induce a form of paralysis. This static situation is unfortunate for a number of reasons. First, it is becoming less and less likely that the third LOS Conference, even if it should produce positive results, will be convened in 1973 as tentatively scheduled. The snail's pace of preparatory work suggests at least the possibility of its postponement. Secondly, as already noted, there appears to be a limited prospect that these negotiations can provide a solution to North Pacific problems.

Meanwhile events do not stand still and new problems in international fishery that continue to develop in the North Pacific will be dealt with in the usual, fragmented way. Although most of the agreements resulting from negotiations from year to year will be temporary in nature, they could set a pattern of negotiations or precedents in the form of allocation which would create further difficulties during their transformation into a new convention and organization.

It is conceivable that by late 1973 or early 1974 we may have a better idea of the outcome of the Law of the Sea Conference on various issues and specifically on questions related to international regulation of fisheries. This is not a long time to wait; the present round of discussions on ocean regimes started in 1967. Moreover, unless our predictions on the outcome of the conference are grossly incorrect, the need for improved regional arrangements in addition to, or in lieu of, a worldwide fishery regime will by then receive much greater attention than it does now. The wait-and-see attitude now taken by nations may have dissipated, and they may again be prepared to consider concrete problems seriously.

Initiative. The question of which nation or nations might take the initial action toward a new agreement is a very touchy one. The question of diplomatic relations is undoubtedly a barrier, although not insurmountable. The North Pacific region is characterized by the lack of formal diplomatic relations among some of the major powers. In view of recent political moves, this picture will probably change substantially as the People's Republic of China establishes diplomatic relations with a widening group of nations. The lack of formal diplomatic relations does not, of course, by itself constitute a legal obstacle to the participation of the states concerned in an international fishery convention, as many precedents demonstrate. But the problem is particularly serious in the North Pacific because of the political volatility of the status of China and Korea. No matter what nation or nations initiate action, the question of whom to invite to participate in initial negotiations is difficult to handle.

The most realistic and desirable combination of promoters consists of the United States, Canada, Japan, and the Soviet Union. Each of the other possible parties might be persuaded to participate through discussions with one or more of the four promoters as political relations suggest.

It is possible that, although all four nations recognize the merits of a new convention, none will be willing to take the lead. In such an event, FAO might be able to perform an important function in the process of formulating a North Pacific fishery agency. FAO could be requested to serve as a promoter as it did in the case of two recent international fishery conventions, namely the International Convention for the Conservation of Atlantic Tuna and the Convention on the Conservation of the Living Resources of the Southeast Atlantic.

Negotiating a new multilateral fishery convention is a time-consuming process. Even with a substantial amount of preparatory work, a period of two to five years is likely to elapse between the initiation of formal negotiations and the coming into force of a new convention. In the interim, the existing agreements and new short-term agreements (most of them probably bilateral) would have to resolve immediate problems. The pattern of negotiations of these short-term arrangements would vary greatly, depending on whether the major fishing nations in the region were in the process of negotiating a new comprehensive convention on the basis of a well-prepared draft or drafts. Each party would have in mind what the others are proposing for inclusion in the new convention. Certain guidelines or principles might be developed as negotiations proceed. Also, once the convention was concluded, it could have substantial effects on interim arrangements even before it came into force.

Membership. For an effective convention, the initial membership must include at least three nations—Japan, the Soviet Union, and the United States. Canada's participation is highly desirable to increase the effectiveness of the new treaty, since Canada has an interest in most of the problems to be faced from the beginning by the new agency: salmon, halibut and other groundfish, herring, and fur seal. The inclusion of South Korea as an initial member would greatly enhance the effectiveness of the agency and might incline North Korea towards participation. In any case, it is important that the new treaty be open to adherence by any party interested in exploitation of North Pacific living resources. Although a comprehensive agreement limited to the initial parties would still be an improvement over the present system, one of its most serious deficiencies would remain: the inability to resolve problems of new entrants into a fishery.

Because the new convention would, preferably, represent a substantial departure from the past, membership should be irrevocable for a period of ten years at least, permitting the parties to rely on the expectations created in each other without threatening harm from a sudden change in interest. After the initial period, it is customary to permit withdrawal upon a year's notice. The agency established by the proposed treaty should survive the termination of membership by a major state, whether voluntary or otherwise, but withdrawal by two such states probably ought to be enough to terminate the agency.

Geographic Area Covered. Two principal problems involve the area to which the new convention applies: (a) the part of the North Pacific to which the treaty applies as a whole, and (b) the waters within which coastal state regulations are applicable.

Care should be exercised with respect to the general coverage of the treaty because membership is open to all interested in the area. For this reason, it is best not to define the area too extensively. The major alternatives in definition appear to be the Pacific north of the equator or north of the Tropic of Cancer. Our preference is for the latter, since it obviates the potential participation of a number of Southeast Asian nations in the initial negotiations, thus avoiding additional complications.

Definition of the treaty area does not determine participation in the agency, since any nation that has developed a substantial fishing interest in the area would be entitled to accede to the treaty. However, unless the present pattern of fishing or the political situation changes greatly before initial negotiations, the suggested area definition reduces potential participants to a relatively few political bodies. Exclusion of the Southeast Asian states from this arrangement, and the expectation that their problems will be dealt with in a separate treaty, seem reasonable because their fisheries problems are so vastly different from those of the North Pacific. Solution of North Pacific problems is difficult enough without adding the tangled affairs of states bordering the South China Sea.

The second principal problem concerns the possible exclusion of each party's territorial sea or fishery zone from the coverage of the treaty, a not uncommon practice. Although it seems unlikely, it would be desirable if an agreement could be reached on the limit of the territorial sea or fishing zone. In the absence of such an agreement, the usual acceptable formula is a caveat to the effect that nothing in the treaty affects the rights, claims, or views of any contracting party in regard to the limits of the territorial sea or of coastal jurisdiction over fisheries. It must be emphasized, however, that in this instance this device is neither recognition of a convenient escape from the treaty's provisions nor a concession to any unilateral competence to change boundaries. The purpose is rather to acknowledge the absence of agreement on a critical question and, by additional provision, to establish a mechanism for coping with possible or actual changes in these limits. We are convinced that the new agreement must at least be able to accommodate negotiations, multilateral or bilateral, between members concerning problems arising from the extension of fishery jurisdiction.

A variety of ways might be employed to approach this difficult question. For example, assuming that agreement on this issue is not reached at the LOS conference, the initial members of the new convention might agree to recognize a *minimum* 12-mile zone (either as a territorial sea or a combination of a territorial sea and an exclusive fishing zone). Then, the convention might provide further that, should a member state subsequently claim further extension of fishery jurisdiction, the matter would be discussed within the agency to examine the justifications used, its impact on the existing arrangements under the convention, and possible measures to compensate for the adverse effects of the particular action on the fisheries of some of the other member states. Such examination would be voluntary on all sides, with the express proviso that it did not constitute recognition of the lawfulness of the claim made. This procedure would in effect accommodate bilateral negotiations within the framework of a regional convention, and make it possible to examine the particular problem in light of its possible effects on the overall management system of the North Pacific.

Objectives

Preambles are commonly used to express the broadest formulation of the goals shared by the parties while somewhat more specific articulation is placed in the treaty text itself. The purpose of the North Pacific fisheries agency can be summarized by such very broad wording as "rational management of fishery resources," a concept which would maximize flexibility. Somewhat greater specificity, however, might avoid or alleviate unnecessary controversy. This might be accomplished by stating the general purpose as, for example, the maximization of yields from fishery resources in the convention area, and, at the same time, minimization of international disputes over fishery matters by accommodating to the greatest extent possible the national interests of the contracting parties in the exploitation of fishery resources.

These latter two objectives are not necessarily compatible with each other. Perhaps it would be more felicitous to state that the convention aims at maintaining yields at a high level, while giving due consideration to the need to accommodate the interests of the parties in exploitation of such resources.

These broad general statements on goals in the preamble should be supplemented in the text by more explicit directions. Among these, it might be useful to refer to the following aims: first, of providing a structure for decision designed to facilitate the management of fisheries in the North Pacific Ocean; second, of establishing a mechanism for acquiring the data needed for effective, timely management decisions; and third, of providing for allocation in an equitable fashion in such forms as seems suitable to the parties concerned. The third goal is an essential component of a new agreement and its formulation is significant. As phrased above, the term "allocation" is unqualified by reference to the things to be allocated. The modifying phrase "in such forms as seems suitable to the parties concerned" emphasizes that the intent is to allow flexibility in choice on this point. Thus, if a catch limit is applied to a stock or stocks, the parties concerned might prefer, in some cases, only to set a total quota. In other instances, however, they may wish to implement a system of allocation in one or another of a variety of forms.

Methods for Achieving Objectives

In view of the changing nature of fishing operations, it does not seem appropriate to spell out in convention articles the specific measures to be taken for conservation purposes; namely, how to maximize yields from the resources concerned or to minimize damaging effects thereon. The agency should be authorized to recommend any measures that it considers most effective for solving specific problems.

However, some discussion of methods to meet problems arising from allocation is appropriate. A new convention would have to face two problems in this respect: first, the need to accommodate the existing arrangements for allocation under the new regime, and second, the development of general forms of allocation to be applied in the future. The first problem has been discussed under "Transformation of Existing Arrangements" at the beginning of this chapter and will be considered again in the context of the structure of the new agency. The second has two components: general forms of allocation in international waters and extension of national jurisdiction as a form of allocation. Here we discuss only the first of these; the second has been treated to some extent in previous discussion under "Geographic Area" and will be examined further in connection with the structure of the agency.

The convention text should not only mention allocation as a general goal but should also refer to certain available forms, including national quotas, allocation of fishing areas, and sharing of benefits. At the same time, the choice of form should be left open, and the agency should be authorized to recommend specific types of allocation in specified situations. Since many large North Pacific fisheries are not now regulated internationally, the arrangements likely to be developed can draw upon a wide range of experience without being bound by prior practices in the particular fishery.

At the lowest level of sophistication in allocation practice, a simple national quota system is likely to be the most acceptable form of allocation for the North Pacific states. Criteria for establishing national quotas might vary from one instance to another but would probably be based upon a combination of factors, including historical interests, measured in terms of the actual catches from a resource or resources over a period of time, the capability of harvesting the amount allocated, and appropriate indicators of the special interest of coastal states.

As a general principle (with exceptions in specific cases), a portion of the total catch limit set for conservation purposes should be reserved as an open quota to be taken by existing and new participants over and above the quotas otherwise allotted to them. In order to make the national quota system workable, the open (or free competition) part of the quota should be fairly substantial; and reviews should be made periodically so that national quotas could be revised in light of actual achievements. This system of allocation is feasible under the domestic institutions of all the potential members of the proposed new agency.

From an economic perspective, it is desirable that national quotas, or part of them, be made transferable in order to maintain higher economic yields. Such a system has a precedent in the Antarctic whaling business, although the arrangements are made outside the Whaling Commission itself. This alternative merits serious examination, although it seems unlikely that it will be acceptable as a general principle to the probable members of the new agency.

Under the national quota system, each national participant would decide how to harvest its quota. A nation might restrict the number (and/or type) of vessels, or it might wish to create subquotas assigned to individual fishing vessels or mother-ship fleets. On the other hand, a nation might wish to harvest its quota as quickly as possible in order to move the fleet to other areas for the rest of the season. Others might merely allow their fishermen to continue fishing in unregulated numbers until the season is closed by the fulfillment of the quota. The use of different harvesting strategies under the various quotas would create technical problems; but these could be resolved through negotiations since, in this region, the number of nations involved in harvesting the same resource is likely to be relatively small.

At another and higher level of sophistication, the nations might agree to a regulatory system that limits not only the amount of fish taken by each, but also the amount and kind of fishing effort (the latter in terms of the number and size of vessels permitted to fish) in order to attain some uniformity in fishing strategy. Certain types of gear might also be banned for this same purpose. Such a system of limited entry, as desirable as it is for improved management, will be difficult to adopt as a general form of allocation because the interests of North Pacific nations in particular resources vary vastly. Various measures to limit fishing effort under a national quota system should not, however, be excluded from possible forms of allocation. Should the nations directly concerned be willing to adopt such measures, the new convention should be able to accommodate them, for direct control of fishing effort is implemented under many of the existing arrangements on the Asian side, either as part of an official agreement or as a voluntary act.

Structure of the Agency

Commission. Generally international fishery management agencies have one body, commonly called the commission, which is the principal authority within the entity and which exercises the most general competence conferred upon the agency by its creators. The treaty would also contain the customary provision that commissioners may be accompanied by such experts and advisers as are needed.

Executive Director and Staff. Most conventions establishing fishery agencies contain only very general provisions for appointment of the principal staff officer and offer surprisingly little direction as to his functions. For reasons elaborated upon in connection with the research function of the agency, the director should be given much greater responsibility than the usual administrative work and should have a staff of at least five or six professionals to execute the tasks involved. The major functions to be discharged by the staff should be defined in the convention and would include the coordination of national research programs, the collection and analysis of statistical data, appraisal of research programs for adequacy of content and direction, and review of research problems and findings referred to it.

The proposal thus suggested is a compromise between employing a full-blown research staff and confining the staff solely to routine duties. We do not doubt that an independent research staff, equipped with sufficient personnel and facilities to do its job, would be a prime means for coping with fishery management problems in the region; and such an arrangement is a first desideratum. We are aware, however, of the several pressures that make a staff operation of this scale and complexity very difficult and probably impossible to realize at this time. At the same time, there is a definite need to ensure that the speed and timing of scientific work are adequate to cope with the very complex problems that would confront a comprehensive management system of the kind recommended here. The relatively modest professional staff could prove invaluable in this sense and might not attract strong opposition. The staff should also be authorized to hire consultants from time to time, depending on the research problem to be examined. This not only bolsters the expertise of the staff but has the additional virtue of helping maintain the objectivity of the staff's judgment.

Decentralization of Authority Within the Agency. A method for distributing authority over various areas and species is an indispensable approach to an improved institutional structure in the North Pacific. For this purpose a panel system resembling, but not completely following, the International Commission for the Northwest Atlantic Fisheries (ICNAF) model is recommended. The vastness of the areas concerned, the small number of parties involved, the complexity of present regulatory and allocation schemes, and the need to transform them in the context of developing a coordinated and comprehensive system, all suggest that a combination of area and species panels is needed. An area panel system alone would be ineffective because it could not cope with the transformation question. Area panels are also needed, but their coverage will have to be much larger than that of the ICNAF areas; otherwise there would be several areas in which only one contracting party had a fishing interest.

The combined area and species panel system must have built into it a substantial degree of flexibility so that it can be modified as needed to reflect changes in the pattern of international fishing activities. The specifics of panel structure and coverage are best dealt with in annexes to the convention rather than in the main text, since the dual structures of the panels and the problems involved will probably entail revision more frequently than is feasible in the basic treaty itself.

Two annexes could be employed to define the structure and work of the panels. Annex I would include the definition of the species or area coverage of the panels. Annex II would consist of a schedule incorporating the various measures of transformation as discussed above and such other relevant measures as the parties believe are required in the transition process. Both annexes could be amended as necessary to reflect changes in panel structure and regulatory measures. Based on the analysis in the discussion of transformation, the panel structure might be as follows:

Species panel 1 (salmon): *Oncorhynchus nerka, O. gorbuscha, O. keta, O. kisutch, O. tachawytscha, O. masou,* and steelhead *(Salmo gairdneri).*

Species panel 2 (crab): *Paralithodes camtschatica, P. platypus, P. brevipes, Chionoecetes* spp., *Lithodes aequispina, Erimacrus isenbeckii.*

Species panel 3 (herring): *Clupea harengus pallasi.*

Species panel 4 (fur seal): *Callorhinus ursinus.*

Area panel 1: The East China Sea, including the Yellow Sea.

Area panel 2: The Japan Sea.

Area panel 3: The Okhotsk Sea and the waters around the Kurile Islands. Waters along the east coast of Kamchatka, outside the Bering Sea, might also be covered by this panel.

Area panel 4: The Bering Sea and the Aleutian waters.

Area panel 5: The western half (west of 180° E) of the Pacific Ocean proper.

Area panel 6: The eastern half (east of 180° E) of the Pacific Ocean proper.

Because the problems involved and purposes served are different, area designations for catch and effort statistics should be established separately from the area division for panels. It would appear appropriate to establish more than one set of statistical areas. For example, statistical areas for salmon might well differ from those for groundfish. No statistical areas are required for fur seal as long as pelagic sealing is prohibited. In any case, the whole question of statistics, including the possibility of setting up a data center, should be studied intensively, before the commission is established, by a group of special research workers from the nations concerned, joined by consultants from other organizations as required.

Annex II should be closely related to the panel structure outlined above, and should spell out the particular measures for conservation and allocation agreed upon by the contracting parties by the time of signatures of the convention. In order to avoid the need for revising the entire Annex II each time changes are introduced, it might be divided into several parts. The following are suggestions for different parts of the annex:

Part 1 (salmon). For the reasons mentioned in detail in "Existing Arrangements and Ways to Transform Them," the most practical course of action concerning salmon would be to start with maintenance of the status quo, and gradually move to better management systems. Should this basic philosophy be acceptable to the initial contracting parties, this part of Annex II would incorporate the substance of salmon regulations under the Japan-Soviet Union Convention Concerning the High Seas Fisheries of the Northwest Pacific Ocean, which would have to be amended year after year. It would also prohibit offshore fishing for salmon in waters east of 175° W except for fishing by the coastal states (in practice troll fishing by Canadian and American fishermen). For obvious reasons, the word "abstention" should be avoided; and it would further be advisable to define the areas in which Canadian and American troll fishing are to be allowed. This procedure would make the salmon part of Annex II less awkward than it would be with a complete lack of international restrictions on high-seas fishing by American and Canadian fishermen, but the practical effects would be the same as the present situation. The same approach should be employed if USSR fishermen begin high-seas salmon fishing. Part 1 should also include a provision to the effect that a member nation with no historical interest in high-seas fishing in the designated areas (which would include the current regulatory areas of the Japan-Soviet Union treaty, waters east of 175° W, and the waters in which Canadian and American fishing are to be allowed) shall not engage in fishing for salmon in such areas. All areas other than those specified in this part would also be closed to high-seas fishing by any party. This would be a de facto prohibition of new entry to high-seas salmon fishing. The year-by-year changes to be made in this part would reflect, practically speaking, the outcome of annual negotiations between the Soviet Union and Japan, except that both Canada and the United States might participate in the sessions of this panel in view of their interest in salmon fishing. If a nation intending to enter high-seas salmon fishing does not join the convention, the matter would have to be settled by political negotiations between the governments concerned.

Part 2 (crab). This part would reflect the existing arrangements for the species concerned between Japan, the United States, and the Soviet Union, now established by three separate agreements, and would be subject to periodical amendment according to the outcome of negotiations among these nations. It might be necessary to include a statement to the effect that nothing in this part of Annex II shall prejudice claims to sovereign rights to explore and exploit the resources of the continental shelf. It would neither be necessary nor desirable to include a provision that would, in effect, exclude new entrants. Should a new nation show interest in participating in the exploitation of the resources concerned, the matter could be discussed by the panel. Problems of crab fishing are not as inflexible as those of salmon, hence some adjustment might be expected.

Part 3 (herring). Initially, any arrangements in effect when the convention was concluded could be incorporated in this part of Annex II. It is expected that these would be amended rather frequently in subsequent years. Canada might or might not insist on inclusion of de facto abstention. The matter could be settled by the establishment of a relatively small closed area for fishing by noncoastal states.

Part 4 (fur seal). This part would incorporate the existing arrangements under the fur seal convention, except for the question of participation by other nations in the sharing system. A provision to reflect the suggestions made in this chapter under the head, "Existing Arrangements and

Ways to Transform Them" (investment requirement) might be considered for inclusion in this part to cope with the problem, although the need for this is not great.

Other parts. The subsequent sections of Annex II would incorporate some of the existing arrangements under bilateral agreements (such as United States-Japan, United States-Soviet Union, Canada-Soviet Union or United States-Canada). The halibut issue could also be dealt with here, although we have no suggestions that might be acceptable to all contracting parties. The issue could possibly be resolved on the basis of preferential allocation to coastal fisheries. It is likely that problems far more important than that of halibut will have developed by the time the convention is signed (for example, pollack regulations).

Any new regulatory measures that might by then have been agreed upon by the initial members could be incorporated into appropriate parts of Annex II. Thus, details in each part of Annex II might be substantially different from those in a new convention if it were signed today, for they would reflect, to a large extent, arrangements in effect at the time that the convention was concluded.

Having determined that the panel system is needed, the key question, particularly in regard to the timeliness of actions recommended by panels, is that of the relationship between the panels and the full commission. Many international regulatory measures, for either conservation or allocation, must be put into effect immediately. If the panel recommendations, particularly those from the species panels, had to be approved by the full commission, which might meet only once or twice a year, it could be difficult to put regulations into force in time for the fishing season of the species concerned. This problem can be solved by giving each panel a substantial degree of autonomy, to be exercised in accordance with conditions established by the commission. First, each panel should be authorized to meet at any time at the discretion of its chairman or upon the request of a member or members of the panel. Second, the panel's chairman should be authorized to forward through the executive director directly to the parties concerned certain panel recommendations, particularly those on catch limits, national quotas, closed seasons, and similar important but changeable measures. Unless the parties notified the executive director of their rejection within a specified period, the recommendations would take effect at a specified time. The commission would have to develop a policy on what types of regulatory measures needed to be treated in this fashion.

There may be other ways of coping with the problem, but one thing is clear: unless the new commission is structured to facilitate expeditious action, the nations concerned will prefer existing arrangements over the new comprehensive system. In spite of all their shortcomings, most existing arrangements have one common asset: actions are taken quickly. Some of the decisions, for example, made by the Japan-Soviet Northwest Pacific Fisheries Commission may be put into effect within a couple of weeks. Most of the decisions taken under bilateral, executive agreements come into force very quickly. In general, too, speed in effective action is a basic requirement for future international fishery organizations in view of the rapid rate at which a major fishery develops. The slowness of existing fishery commissions in taking action is often used by some nations as a main reason for preferring international fishery regimes with main emphasis on jurisdictional control by coastal states.

Committees. The new agency might utilize three permanent committees: one on research and statistics; one on administration and finance; and a special committee to deal with problems arising from claims for extension of national jurisdiction. The question of establishing a committee on research and statistics will be dealt with later under "Functions: Research." If a committee on administration and finance should be needed, it could be established under the commission's rules and procedures and needs no discussion here. This leaves the question of the establishment of a special committee on extension of national jurisdiction. Such a committee is virtually indispensable if the commission is to function as a comprehensive fishery management agency for the North Pacific. Problems arising from jurisdictional claims could be handled by the commission itself, especially in view of the highly political nature of the problems, but a special committee might be more useful and effective. In any case, without this responsibility, the scope of commission work would be much narrower; and members would develop a number of bilateral agreements to cope with these problems outside the commission, thus defeating the very purpose of having an overall regional agreement.

Voting Practices

The degree of independent authority to be granted an international fishery organization is always controversial. It does not seem realistic at this stage that the commission should be authorized to act by a mere majority or even by a two-thirds majority, since the initial membership will probably consist of only four, and possibly five, states. If a two-thirds majority were adopted at the beginning, many commission recommendations would be rejected by the parties and the commission would look rather weak. Moreover, we do not expect a unanimity requirement to be especially burdensome if the initial parties are agreeable on principles of conservation and allocation as described earlier.

Assets

Two major problems are involved in financing any international organization: the total budget and the division of the total burden among the participants. The budget for the agency outlined here would be rather large, since the commission would have a substantial staff for research planning, coordination, and evaluation, as well as collection and analysis of statistics. Support should also be available for contracts for data storage and retrieval (unless some national agency could provide this service free of charge, which is a possibility), travel, publications, and other miscellaneous expenditures. Such a budget would still be far smaller than that of an international fishery body of this scope with its own research staff.

The system of cost sharing to be adopted by the convention would depend on the size of the budget and the number of the initial (and also potential) contracting parties. If the number of parties is very small and the budget is not large, equal sharing is the simplest way to handle the matter. A sharing system based on panel memberships (similar to that used by ICNAF) might be considered if the number of participants is large and the membership differs greatly from panel to panel. Another basis could be the relative amounts (or values) of fish taken from a resource or resources with which the convention is concerned, thus resembling the Inter-American Tropical Tuna Commission. As the new North Pacific convention would probably involve a relatively small number of participants, but deal with an extremely wide range of problems under a complicated panel structure, none of the above-mentioned sharing systems seems particularly suitable for meeting the costs. Perhaps a combination of different systems would have to be adopted.

Functions

Research. Because of difficulties in funding and probable duplication with national research programs aimed at each state's own needs, it is impractical at this stage to suggest the creation of a full-scale research staff in a new all-embracing North Pacific fishery management agency.

On this assumption, there will be at least two requirements for making the system work: (a) machinery for developing well-coordinated programs to be executed by national agencies and for appraising their implementation, and (b) means for expediting agreement among scientists. The first requirement could be satisfied by providing an international staff of moderate size whose functions would include program coordination and assessment of investigations.

The problem of expediting scientific work is more complex and difficult to resolve. As is widely recognized, difficulties in international fishery negotiations are often caused by a lack of agreement on the condition of the stocks concerned and the need for taking regulatory measures. Negotiating parties sometimes demand watertight scientific proof, which is almost impossible to obtain. The problem is intensified by the fact that research results are used as a basis not only for taking conservation measures, but also for implementing allocation in one form or another. We believe, however, that substantial improvement over the present state is feasible.

Initially, nations must recognize that fishery science is still imperfect. Even the best mathematical models so far developed are not widely applicable to interpret the population dynamics of different stocks. The stock-recruitment relationships (a key question for most resources), as well as causes of large fluctuations in the abundance of coastal pelagic resources, are still largely unknown. Also, the exploitation by a fishery of a large resource can easily reach its maximum level within a very few years. Thus, management problems often develop much faster than concrete evidence can be obtained from time-consuming investigations. In some cases, lacking other evidence, predictions on future changes in exploited stocks have to be based on what has happened to similar stocks in other areas under similar circumstances. The new convention would perhaps have to include a provision to the effect that measures of certain types should be taken on the basis of the best evidence available then. In case of controversy, settlement would be based on the preponderance of the evidence.

Another aspect of the process of decision involving scientific matters could be improved. We have already suggested that decisions by majority vote on commission or panel recommendations are probably not practical. This does not create too much difficulty because political considerations are brought into the debate at the commission level in any case. By the time the commission reaches a decision, the political problems are settled; and experience demonstrates that most commission recommendations are accepted by member governments. The performance of many of the existing organizations has been hampered by a lack of separation between scientific judgments and political decisions. The new convention could provide a partial solution to this vexing problem.

For example, a group of specialists might be created to make recommendations, upon request by the agency, on questions of a scientific nature. The group would not arbitrate issues with political implications, but would be restricted to passing judgment, again based on the preponderance of the evidence, on such questions as effects of fishing, the level of maximum sustainable yield (if there is one), or the distribution and migration of a particular population. It might also suggest, if requested, specific conservation measures to be considered, as well as improvements in the research programs carried out under the auspices of the agency. The group would be established on an ad hoc basis and would consist of scientists from nations not directly concerned with the exploitation of the resource in question. While its power would be limited, the group would serve a very useful purpose; that is, to expedite international fishery negotiations and facilitate timely actions.

Another approach would be to give this function to a standing committee on research and statistics, the work of which would cut across all panels. The commission could benefit by having research problems of different panels examined by a single committee consisting of one or two scientists chosen by each contracting party. Reseach problems would first be discussed in an appropriate panel or panels and then referred to the committee on research and statistics for recommendations and suggestions. This committee would also be responsible for all matters concerning statistics.

A serious technical problem could arise in arranging for the full research and statistics committee to meet often enough to examine all matters referred to it by the panels. A compromise, which is not really satisfactory, might be that through the executive director the panels would forward their conclusions and recommendations on scientific matters to the commission for approval. At the same time, the panels would forward the same conclusions and recommendations, with supporting evidence, to the committee on research and statistics. The full committee would meet during the regular sessions of the commission, as well as in the interim as required. The committee would comment on the conclusions and recommendations forwarded by the panels, which would be taken into account in later decisions made by the respective panels and the commission. The committee would then be operating somewhat independently from the commission and the panels, more or less as an advisory body. The committee's comments or recommendations would not be binding but if the committee consisted of scientists of high caliber and international reputation, it would have a substantial influence on decisions taken by panels and the commission.

Yet another device would be to assign this function to the staff professionals. While the staff's comments on the matters referred by the panels need not bind the latter, they could have substantial weight, depending on the caliber of the staff. Between the two alternatives (a committee cutting across all panels and a staff of moderate size to perform the same functions), the latter would be preferable, provided of course that the parties concerned were willing to meet the costs of maintaining such a staff. The commission should still have an option to refer issues of a scientific nature to third-party panels to be formed on an ad hoc basis, as suggested previously.

Authority. We are not proposing that any truly independent body be established to manage North Pacific fisheries since the agency would not have any authority to prescribe regulations. This is not a desirable state of affairs, but we are unable to suggest any benefits to potential members that might be sufficient to induce them to transfer such authority to an international agency. In practice, it is to be hoped that the agency and its main organs would develop sufficient expertise and integrity so that its actions would be tantamount to final decision although in form they were only recommendations.

Recommendations. The proposals made here do contain one major change from the provisions in other conventions, and this concerns the short-cut arrangement for facilitating adoption of panel recommendations. Also, the agency's organs would be authorized to make recommendations on a subject seldom assigned to such bodies, namely that of allocating benefits from the resource concerned.

Enforcement. A system of mutual inspection in international waters should be adopted under the new convention. The authorized officials of a contracting party would be able to inspect a vessel of another party when there were reasonable grounds to believe that the vessel had violated the provisions of the convention in international waters. Any necessary adjudication would occur in the state to which the personnel or vessel belonged. In addition, the convention might instruct the commission to hold discussions with a view to establishing standards for minimum penalties for certain types of violation, which would help reduce their incidence. Otherwise, the system of enforcement might be more or less similar to those now operating under the North Pacific treaty or the Japan-Soviet Union treaty.

SUMMARY

A new, comprehensive arrangement for the international management of North Pacific fisheries is badly needed. Serious disputes are expected to arise, not only over those fisheries currently regulated under a fragmented regime, but also over those now operating in international waters but not subject to international regulations. The latter land a far greater amount of fish than the former. An acceptable overall agreement, however, could not be developed by scrapping the existing arrangements; instead, they should be transformed under the new convention. Except for salmon, this process of transformation does not appear to present insurmountable problems.

The new convention and the agency to be established would have some unique features. The convention must mention allocation (along with conservation) as one of its main objectives. The agency should have a panel structure consisting of a combination of species panels and area panels. To facilitate the implementation of certain types of regulatory measures, the panels would be given power to forward their recommendations concerning such measures directly to the contracting parties concerned. In order to accommodate the existing arrangements under provisions of the new convention, it would have an annex (or a protocol or schedule) reflecting the specific arrangements in effect at the time the convention was concluded. A special committee should be established to discuss effects of new jurisdictional claims with a view to minimizing conflicts from this source.

While the agency need not have a full research staff to carry out all research programs required for the implementation of provisions of the convention, it should have a staff of moderate size, headed by an executive director and consisting of individuals of high caliber with international reputations, to coordinate research programs carried out by different national agencies, to assess the results thereof, and to compile and analyze statistical and biological information. All contracting parties would be responsible for providing catch and effort statistics, as well as certain types of biological information, according to formats to be established by the agency. Substantial financial support would be required largely to support the staff.

At least at the outset of the agency, recommendations would be made unanimously. A system of mutual inspection in international waters would be adopted for enforcement, possibly with the establishment of minimum penalties for certain types of violation.

Because of uncertainties about the outcome of the third LOS Conference, none of the countries concerned appears to be willing to take the initiative for the establishment of a new agency. Meanwhile, disputes over fishery matters in the North Pacific will be dealt with largely through bilateral negotiations, resulting in a further fragmented regime. The possibility of FAO taking the initiative should be considered.

References

AHLSTROM, E.H. 1968. Fishery resources available to California fishermen. In *The future of the fishing industry of the United States,* ed. Gilbert De Witt, pp. 65-80. Publications in Fisheries, new series, no. 4. Seattle: University of Washington.

BURKE, W. 1967. Aspects of internal decision-making in intergovernmental fishery commissions. *Washington Law Review* 43: 115-78.

JAPAN. MINISTRY OF AGRICULTURE AND FORESTRY. 1971. *Yearbook of production statistics for fisheries and agriculture, 1969.* (In Japanese.) Tokyo: Ministry of Agriculture and Forestry.

———. 1972. *Yearbook of production statistics for fisheries and aquiculture, 1970.* (In Japanese.) Tokyo: Ministry of Agriculture and Forestry.

KASAHARA, H. 1961. *Fisheries resources of the North Pacific Ocean.* Part 1. H.R. MacMillan Lectures in Fisheries. Vancouver: University of British Columbia.

———. 1964. *Ibid.* Part 2.

———. 1972. Japanese distant-water fisheries: a review. (National Marine Fisheries Service, National Oceanic Atmospheric Administration, U.S. Department of Commerce) *Fishery Bulletin* 70(2):227-82.

KOBLENTZ-MISHKE, O. J.; VOLKOVINSKY, V.V.; and KABANOVA, J.G. 1970. Plankton primary production of the world ocean. In *Scientific exploration of the South Pacific,* ed. W. S. Wooster, pp. 183-93. Washington, D.C.: National Academy of Sciences.

MATHISEN, O. A., and BEVAN, D.C. 1968. *Some international aspects of Soviet fisheries.* The Social Science Program of the Mershon Center for Education in National Security. Columbus, Ohio: Ohio State University.

MOISEEV, P. A. 1964. Some results of the work of the Bering Sea expedition. In *Soviet Fisheries Investigations in the Northeast Pacific,* ed. P.A. Moiseev, Part 3, pp. 1-21. All-Union Scientific Research Institute of Marine Fisheries and Oceanography; Pacific Scientific Research Institute of Marine Fisheries and Oceanography. Original in Russian; English translation (1968) by the Israel Program for Scientific Translation, available from the U.S. Department of Commerce, Springfield, Va.

SAKIURA, H.; YUHASHI, S.; KOYAMA, Y.; and FURUSE, R. 1964. The fishing industry of the U.S.S.R. *Kaigai Suisan Sosho* (Nihon Suisan Shigen Hogo Kyokai) 4. (In Japanese.)

SWYGARD, K. 1948. The international halibut and sockeye salmon fisheries commissions: A study in international administration. Unpublished doctoral thesis, University of Washington.

UNITED NATIONS, GENERAL ASSEMBLY. 1971. *Report of the Committee on the Peaceful Uses of the Seabed and the Ocean Floor Beyond the Limits of National Jurisdiction.* Official Records: 26th sess., Suppl. no. 21 (A/8421).

U.S. CONGRESS. SENATE. COMMITTEE ON COMMERCE. 1970. *Treaties and other international agreements on oceanographic resources, fisheries and wildlife to which the United States is Party.* 91st Cong., 2d sess.

List of Organizations

BSSSC	Baltic Sea Salmon Standing Committee
CARPAS	Regional Fisheries Advisory Commission for the Southwest Atlantic
CECAF	FAO Fishery Committee for the Eastern Central Atlantic
GFCM	General Fisheries Council for the Mediterranean
IATTC	Inter-American Tropical Tuna Commission
ICCAT	International Commission for the Conservation of Atlantic Tunas
ICNAF	International Commission for the Northwest Atlantic Fisheries
ICSEAF	International Commission for the Southeast Atlantic Fisheries
INPFC	International North Pacific Fisheries Commission
IOFC	Indian Ocean Fisheries Commission
IPFC	Indo-Pacific Fisheries Council
IPHC	International Pacific Halibut Commission
IPSFC	International Pacific Salmon Fisheries Commission
IWC	International Whaling Commission
JKFC	Japan-Republic of Korea Joint Fisheries Commission
JSFC	Japanese-Soviet Fisheries Commission for the Northwest Pacific
MCBSF	Mixed Commission for Black Sea Fisheries
NEAFC	North-East Atlantic Fisheries Commission
NPFSC	North Pacific Fur Seal Commission
PCSP	Permanent Commission of the Conference on the Use and Conservation of the Marine Resources of the South Pacific
SCNEA	Sealing Commission for the Northeast Atlantic
SCSK	Shellfish Commission for the Skagerak-Kattegat

For Product Safety Concerns and Information please contact our EU representative GPSR@taylorandfrancis.com Taylor & Francis Verlag GmbH, Kaufingerstraße 24, 80331 München, Germany